MATH
YOU CAN REALLY USE—
EVERY DAY

David Alan Herzog

Wiley Publishing, Inc.

Library of Congress Cataloging-in-Publication Data:
Herzog, David Alan.
  Math you can really use : every day / David A. Herzog.
      p. cm.
  Includes index.
  ISBN-13: 978-0-470-05402-4 (pbk.)
  ISBN-10: 0-470-05402-6 (pbk.)
  1. Mathematics–Popular works. I. Title.
  QA93.H4 2007
  510–dc22

                         2007005095

Printed in the United States of America

10   9   8   7   6   5   4   3   2   1

Book production by Wiley Publishing, Inc., Composition Services

Wiley Bicentennial Logo: Richard J. Pacifico

This book is dedicated to the memory of my sister, Lois Herzog Cullen, who was taken from us much too soon.

It is also dedicated to Francesco, Sebastian, and Gino Nicholas Bubba, Hailee Foster, Rocio, Myles, Kira, and Reese Herzog, Jakob and Alex Cherry, Marcelo, all their parents, Uncle Ian, and Grandma Birdie.

Acknowledgments

I find it only fitting that I should acknowledge the two Wiley acquisitions editors who have kept me in projects over the past several years, Greg Tubach and Roxane Cerda. Also deserving of mention is Suzanne Snyder, who with this volume has project edited four of my books in a row, and has not only put up with my sometimes weird sense of humor without changing my quips, but has from time to time thrown in or suggested a wry subheading or two. Then there's Christina Stambaugh with whom I have had no direct contact, but who has been a critical part of this whole operation—in that she was instrumental in setting up this book's production schedule and (along with Wiley's designers) in creating an interior design for this book. Finally, I must acknowledge my wife Karen, who has refrained from nagging me to "get a real job."

Table of Contents

Introduction

So, you hate math, and you're not very good at it. Or you hate math *because* you're not very good at it. *Or*, when they taught math at your school you were out that day. Just kidding. I have no intention of telling you that math is good for you, or that this book is going to teach you how to do math and make what seemed hard easy. There are already some good books out there to do just that. I should know. I've written some of them. Rather, this book is meant to give you a practical way to handle the math problems that you can't escape from, as you encounter them in everyday life. Stuff like tipping the waitress (waiter), bell hop, or cab driver, balancing your checkbook, or figuring out how much you'll be saving if you go to a "30% off" sale when the item lists for $124.95. There are many times when a calculator comes in handy, but there are also times when it can be inconvenient or downright embarrassing to have to resort to pulling one out.

This is a friendly book, and what could be more friendly than a book that can save you money? There are some appendices that should come in very handy, since you can photocopy them and take them with you when you need to refer to them at a restaurant, hotel, or department store. Learn how to compare credit card offers, loan offers, and insurance offers. Learn to tell the difference between a good deal on a mortgage and a bad one. You'll get practical help ranging from what to look for in a bank to how to buy carpets, floor coverings, and wall paint. I'll help you convert square yards to square feet or centimeters, and the other way(s) around. The first few chapters will deal with some principles that I'll follow throughout the book. I suggest you read those, just so you're not forced to go, "Huh?" when you pick it up at a later time. After that, go to the chapter that deals with whatever it is you need help dealing with, and hopefully, you'll find the help you're looking for. If you don't, let me know. You just might be providing me with a good subject for writing another book. Needless to say, I'm always on the lookout for one of those.

Close Enough

When we learned math in school we really learned it backward. Our mathematics education—particularly our arithmetic education—began with the least important numbers and worked through the years to get us to the more important ones. I guess that follows the theory of having to learn to walk before you can run and defers to the lower capacity of the young child's mind to grasp more complex models. Historically, mankind probably began to be concerned with numbers when the idea of ownership of livestock began, or maybe even while we were still hunters and gatherers. The first three numbers were probably designated by one finger for 1, two fingers for 2, and a whole hand's worth of fingers for many, to designate how many wooly mammoths were ambling into hunting range. I can't conceive of a need for a more specific greater number than **many**—at least before we became herders and needed to keep track of our sheep and goats.

It's the Big Ones That Count

The fact of the matter is that the larger numbers are more important than the smaller ones. If you were about to buy a car that cost $23,472.87, would you reject the deal if the sales manager said, "Whoops, we have to raise the price to $23,472.89"? I don't think so. On the other hand if that raise of 1 were in the ten thousand dollars column rather than in the pennies column, you might stop and reconsider whether the car is really worth $33,472.87, not to mention whether you can afford it. I've often wondered why they even bother with the five right-hand digits when they make up the price tags of automobiles. I don't think I've ever seen a house for sale with a price tag of $247,362.43. Have you? That house would be priced at $245,000, or thereabouts.

The point is, when you're dealing with large numbers, the exact amount really doesn't matter. Who cares if 47,236 people were at Friday night's baseball game, or if there were really 47,186 fans present? There were *about* 47,200 fans at Friday night's game, and that's what we call a **ballpark figure** (pun intended). The *rounded* attendance number for Friday night's game is 47,200. Likewise, when you're trying to figure out how much of a tip to leave for the waitperson at a restaurant, you don't need to compute it based on a check totaling $29.67. Round the bill to $30 and compute the tip (we'll have much more about that in Chapter 6).

In case you need a reminder about the "places" in place-value numerals, check out the chart and explanation below.

Thousands   Ones

h	t	u	h	t	u

Thousands and Ones are the first two periods. Within each period are the hundreds (h), tens (t), and **units** (u) or ones. So, a **digit** (a single-place numeral) written so as to be positioned below the right-hand "t" would have a value of 10 times that digit's name. A digit written so as to be positioned below the left-hand "t" would have a value of 10,000 times that digit's name. If we moved one more period to the left we would be in the Millions period, and any digit would have a value of its name times the place's heading (h, t, or u) times one million. That's how we're able to represent very large numbers using only the ten digits 0 through 9.

Rounding Up or Down

When rounding numbers, there are two key things to remember:

1. Always look one place to the right of the place you're rounding to. (If you're rounding to tens, look in the ones place; rounding to hundreds, look in the tens place; etc.)
2. The key number is 5: 5 or more rounds up; less than 5 rounds down.

Here's how to apply these rules. To round 47<u>6</u> to the nearer 10, look to the ones place (it's underlined). The number in the ones place is a 6. Because 6 is 5 or greater; you **round up.** So 476 to the nearest ten rounds to 480. That won't be our last round up. This is a concept that can help you through all sorts of mental math calculations.

To round 47<u>2</u>8 to the nearest hundred, look in the tens place (it's underlined). Because 2 is less than 5, we **round down.** Rounding 4728 down to the nearest hundred gives you 4700.

You may have noticed that we didn't use a comma in any of the thousands numbers in the preceding paragraph. But we did when we talked about "ballpark figures." That's because it is now conventional to not place a comma until we get into the 10,000s. If you're more comfortable writing "4,728" feel free to keep doing it your way. The one sure thing is you won't be entering any commas on your calculator.

Pop Quiz

1. Bill needs 339 shingles to repair his roof. Shingles are sold in boxes of 100. How many hundred shingles should Bill buy?

2. Barbara needs to withdraw $15,623 from her brokerage account to pay for a kitchen remodeling job. She is required by the account's terms to withdraw only multiples of $1000. How much does she have to withdraw?

3. Frank figures that he'll need about 550 pounds of concrete for the patio he's putting in. To get the best price on the concrete, he needs to buy it in hundred pound sacks. How many hundred pound sacks should he buy?

4. Reese is buying a slightly used car with a sticker price of $9398. After some haggling, the seller agrees to accept a price rounded to the nearer thousand. What should Reese pay for the car?

Answers

1. If this were just a math problem we'd be looking for the nearest hundred, so the only digit we'd pay attention to is the 3 in the tens place. Because 3 is less than 5, we'd round down to 300, but that would leave us short by 39 shingles. That would let a lot of rain come in. Bill will need to buy 400, and he'll have some spares around for emergencies.

2. Here, we're looking for the nearer thousand dollars. That's the digit to the left of the comma. The only digit we are concerned with is the 6 (in the hundreds place). It is greater than or equal to 5, so we round up to $16,000.

3. Though it seems to be in the middle, 550 rounds up to 600, because the tens digit is 5 or greater. I'm sure Frank will find a use for the extra 50 pounds of concrete.

4. Did you pay any attention to the 9 or 8 in question 4? You shouldn't have. Only the 3 in the hundreds place need concern you. That makes it round down to $9000, so Reese saved some money.

2

A Number of Realms of Numbers

Numbers are divided into many families, or **realms.** Though the following discussion is admittedly academic rather than practical in nature, I think it's pretty cool and you might find the distinctions interesting.

The Realm McCoys

First among the realms are the **natural numbers,** or **counting numbers.** The second name describes what you use them for, to count: 1, 2, 3, 4, . . . are the counting numbers, and they're infinite. That means they go on forever, without end. Counting numbers may be represented as all the whole numbers on a number line that are to the right of zero.

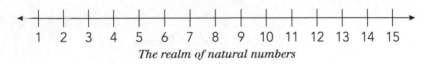
The realm of natural numbers

Counting numbers are also known as **cardinal numbers,** as distinguished from the **ordinal numbers** (first, second, third, and so forth).
The simple addition of 0 to the realm of natural numbers brings us into the realm of **whole numbers,** and we're not done yet.

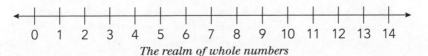
The realm of whole numbers

INTEGERS

As the double arrowheads on both the preceding number lines indicate, the numbers continue in both directions. Once you've crossed 0, you've created a world that contains negative whole numbers (to the left of 0) as well as positive ones (those to the right of 0). Zero is neither positive nor negative. This new realm, which includes the two preceding ones, is that of **integers.** Because the natural numbers without 0 or negatives are infinite, the realm of integers is also infinite, as is the realm of whole numbers.

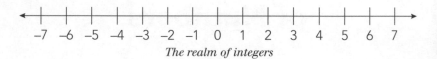

The realm of integers

You probably learned about integers somewhere around junior high school or middle school (essentially two different names for the same place). That was probably also the first time you asked yourself "What in the world do I need this stuff for?" Well, you may not have found a use for algebra and trigonometry, and especially calculus, but integers are very important in everyday life. What happens if you have $10 in your checking account and you write a check for $25? You've spent $15 more than you actually had. Your checking account balance would now be –$15 (read that negative fifteen dollars).

What if the temperature outside is 5 degrees and it's scheduled to drop tonight by 30 degrees? What will the temperature be then? I trust you figured out that it would be 25 below zero, also known as negative 25 degrees (–25°). So as these demonstrations show, the

The Origins of Zero

Zero was actually brought into use in Europe by the Italian mathematician Leonardo Fibonacci, who grew up in Arab North Africa, around 1200 C.E. Up until then, all commerce in Europe was being done with Roman numerals—a considerable inconvenience. While the Arabs are credited with having introduced Europe to the concept of 0, as well as the current set of numerals we use, they got it from the Hindus sometime in the seventh century. Interesting as that may be, you try multiplying or dividing with Roman numerals and see how far you get.

realm of integers is a very useful one, especially when it comes to weather and finance. But there's more.

RATIONAL NUMBERS

Look at any one of the preceding three number lines. What go between those hash marks (also known as *hache marks*)? We remember from geometry that a line segment is a continuous string of points. Shouldn't there be values for the other points on that line, and not just the few that we've marked? Indeed there are. In order to reach them we must once more expand the realm, to now include **rational numbers.**

A rational number is any number that can be expressed as a fraction, such as $\frac{1}{2}$, $\frac{1}{3}$, $\frac{3}{5}$, or $\frac{5}{8}$. Any integer can be written as a fraction. For example, 6 is equal to $\frac{6}{1}$ or $\frac{12}{2}$. A common fraction has two parts, as illustrated below:

$$\frac{\text{numerator}}{\text{denominator}}$$

Consider a pizza pie that has been cut into eight equal slices, three of which are still in the pan. The denominator tells us into how many parts the whole has been cut. The numerator specifies how many of those slices are being considered. So the number of slices of pizza still in the pan compared to the original intact pie is represented by the fraction $\frac{3}{8}$. The number of slices missing from the pan is represented by the fraction $\frac{5}{8}$. When the pie came out of the oven and was first sliced, there were $\frac{8}{8}$, which is equivalent to one whole pie.

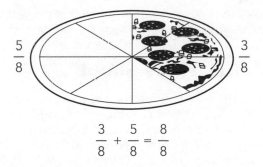

$$\frac{3}{8} + \frac{5}{8} = \frac{8}{8}$$

Not only is the realm of rational numbers infinite, but the number of rational numbers that may exist between any two consecutive whole numbers is also infinite. That is to say, there are an infinite number of fractions between 1 and 2, an infinite number of fractions between 2 and 3, and so on.

Any common fraction may also be expressed as a decimal or a percent. Certain common fractions, when expressed as a decimal, form repeating decimals, for instance $\frac{1}{3}$ = 0.333333333 . . . ad infinitum. Also, $\frac{1}{6}$ = 0.1666666666. . . . Repeating decimal fractions are considered to be part of the realm of rational numbers.

IRRATIONAL NUMBERS

Other fractions form decimals that do not follow a pattern of repetition. The fraction represented by the Greek letter pi = $\frac{22}{7}$, which forms the decimal fraction 3.142857143 . . . Computers have figured out the value of π to thousands of decimal places with no discernible pattern of repetition found. Another such number is the square root of 2 ($\sqrt{2}$), which equals 1.414213562 . . . Again, no recognizable pattern of repetition. Both π and $\sqrt{2}$ belong to the realm of **irrational numbers.** This might really try your credulity, but the realm of irrational numbers is also infinite.

When combined, the realm of irrational numbers plus the realm of rational numbers forms the realm of **real numbers.**

There is also a realm of numbers you may remember having come across in high school. That is the realm of imaginary numbers, and they look like this:

Imaginary numbers take the form $a + bi$, where i is the square root of negative one.

You'll be pleased (and relieved) to know that this is the last time in this volume that imaginary numbers will be mentioned.

Once, Twice, Three Shoot

The realm of integers may be broken into categories other than positive and negative. There are **even** numbers and **odd** numbers (but not strange ones). I named even before odd, because odd numbers are usually defined in terms of evens. An even number is any number that is a multiple of 2, that is, any number that ends with a 0, 2, 4, 6, or 8 in its ones place. All the rest are odd. If you'd prefer, you could say odd numbers are ones that end with 1, 3, 5, 7, or 9 in their ones place. All the rest are even. Even numbers are those perfectly divisible by 2. Divide an odd number by 2 and there will always be a remainder.

When I Was in My Prime

When numbers are multiplied together to form other numbers, those numbers being multiplied together are known as **factors.** In the expression $2 \times 3 = 6$, 2 and 3 are factors of 6. Certain special numbers are called **prime numbers.** Each prime is a number that has exactly two factors, itself and 1. Considering natural numbers, the first prime is 2. Do you know why 1 is not a prime? Because 1 has exactly one factor, not two. You may think about that for a few seconds if you'd like.

Our first prime, 2, has exactly two factors, itself and $1:2 \times 1 = 2$. That's it. Reversing it and writing $1 \times 2 = 2$ is just another way of saying the same thing. Not only is 2 the first prime; it is also the only even prime. If you think back to the first definition of even, you'll realize why. Every other even number will have at least three factors, with 2 being one of them. Can you find the next four prime numbers? Take your time, I'll wait. Got 'em? They're 3, 5, 7, and 11. They came pretty close together, didn't they? The next three come pretty fast too. They're 13, 17, and 19. Can you find the next three? You'll find that they don't come along quite so fast. Not 20, it's even, so it can't be prime. And 21 has factors of 3 and 7. You'll finally hit the next prime at 23. After that comes 29 and then 31. While there were eight primes in the first twenty natural numbers, there are only four in the next twenty. And you'll find that the higher you go, the fewer they'll become. Still, the quantity of prime numbers is, once again, infinite.

I told you earlier that this was an academic exercise, but in reality, prime numbers can be very useful when it comes to working with fractions, as you'll see later.

Pop Quiz

1. The current temperature outside is 18°C. It is expected to drop by 30° after the sun goes down. Using no words, express what the temperature will be at that time?

2. a) Write the cardinal number that describes the number of eggs in 2 dozen.

   b) Write the ordinal number that names the last of those eggs.

3. At Hailee's birthday party, members of her mother's family and members of her father's family were in attendance. Her birthday cake was cut into twenty-four equally sized pieces, fourteen of which were consumed by her father's family members. Assuming that her mother's family ate the rest of the cake, write the fraction that expresses the portion her mother's family ate in lowest terms.

Answers

1. –12°C (read negative 12 degrees)
2. a) 24
   b) 24th
3. Her mother's family ate the remainder, $^{10}/_{24}$, which in lowest terms is $^{5}/_{12}$. Alas, poor Hailee didn't get any cake.

Arithmetic of Whole Numbers

This is a chapter that I didn't think I was going to write, until I realized that depending on tricks to do everything is really not the way to go through life. In coming chapters I'll provide you with tip cards and tricks to use for mentally manipulating numbers, but I'm adding this refresher on arithmetic just because you may have been away from it long enough to have forgotten much, if not all, of what you once knew. Don't feel obligated to read this chapter. If you already know all of the stuff in it, you can skip it if you want.

There are basically two major types of arithmetic operations. The first type is a **combining operation,** whereby two or more things are put together. The combining operations are addition and multiplication. The second type is the **taking apart,** or **undoing operation.** The undoing operations are subtraction and division. The two combining operations are closely related to each other, as are the two undoing ones.

> Lewis Carroll's Mock Turtle was known to refer to the "different branches of Arithmetic [as] Ambition, Distraction, Uglification and Derision."

Addition

Addition is an operation you can do mentally or on paper. Though you might find a calculator easier if you're adding a really long column of numbers, mental addition is really pretty easy. Suppose you needed to add 2353 with 5428. I know you were taught to start on the right and work your way left, and if you were going to do it on paper I'd concur, but mentally, start with the thousands.

$$2353$$
$$+\,5428$$

Starting at the left, 2000 + 5000 = 7000. Go on to the hundreds: 300 + 400 = 700. That's 7700 so far.

Next, move on to the tens, 50 + 20 = 70, so we have 7770.

Finally, 8 + 3 = 11, so add 11 to that and you end up with 7781.

Now I'm making one big assumption here. I'm assuming that you know your addition facts through 18. If you don't (and you know who you are), take out some flash cards and learn them. We have included some at the back of the book, for you to cut out. There are only 45 addition facts. The **commutative property** for addition helps keep the number down. It's just a fancy way of saying order doesn't matter: 5 + 3 = 3 + 5 and either combination makes 8. So, in the flash cards at the back of the book, you'll find 3 + 5. But you won't find 5 + 3. Why learn the same fact twice? If it helps, though, you can read it backward and forward.

It is only possible to add two numbers at a time. The **associative property** for addition states that when adding 3 or more numbers, the way you group them does not affect the result. When adding 5 + 3 + 4, you may first add 5 + 3 and then add the 4, add 5 + 4 and then add the 3, or add 3 + 4 and then add the 5:

$$(5 + 3) + 4 = (5 + 4) + 3 = (3 + 4) + 5 = 12$$

Column addition is modeled below. If the columns are ragged (not all the same number of digits on each line) line up the right hand digits, so all others fall into place.

To add 345 + 2316 + 284 + 97, first set them up like this:

$$345$$
$$2316$$
$$284$$
$$\underline{97}$$

Look for combinations that make 10. That makes the addition easier. So first add 6 and 4 to make 10, then 5 + 7 = 12, and 10 + 12 = 22. We add 10 to any number just by bumping up that number's tens digit by one, so 12 becomes 22. Now write the 2 in the ones column, and rename the 20 as 2 tens, like so:

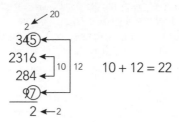

Next, turn your attention to the tens column. There's a 9 and a 1 and a 2 and an 8. That's two tens, plus 4, to make 24 (tens understood), so put the 4 into the tens column and rename the 20 tens as 200; a 2 in the hundreds column.

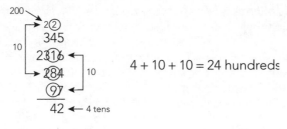

The hundreds column contains two 2s and two 3s. That's $6 + 4 = 10$ hundreds, for which we write a 0 in the hundreds column, and bring the one thousand (10 hundreds is 1 thousand) into that column.

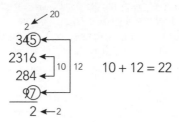

Finally, add the thousands.

$$
\begin{array}{r}
1\,2\,2 \\
345 \\
2316 \\
284 \\
97 \\
\hline
3042
\end{array}
$$

Ta da!

Multiplication

I'm guessing you were expecting subtraction here, because that's how it was taught to you before. I prefer to stick first with combining operations, and the second one of those is multiplication. **Multiplication** is a shortcut for repeated addition: 5×7 is shorthand for $7 + 7 + 7 + 7 + 7$, or 7 added 5 times. Any multiplication can be rewritten as an addition, but that would defeat the purpose of the operation. A shortcut, after all, is a method of doing something faster than it might normally take. Of course, in order to do that, there are some things you need to know.

Just the Facts, Ma'am

Those things that make multiplication a shortcut for addition are called **multiplication facts,** or, I shudder to say it, the "times tables." Now certain times tables come easily to most of us, and some require no thinking whatsoever. The ones table, for example, is the number itself. That is, $1 \times 4 = 4$, $1 \times 7 = 7$, etc. In fact, one is known as the **identity element** for multiplication, because multiplying by it doesn't change the value of the number. (The identity element for addition, by the way, is 0. Can you see why? Adding 0 doesn't change a number's value.) Because multiplication is a form of addition, the associative and commutative properties apply. That is,

Associative: $(2 \times 3) \times 4 = 2 \times (3 \times 4) = 24$

Commutative: $3 \times 2 = 2 \times 3 = 6$

Most people learn the 2s table intuitively even before they are taught to multiply. That's because there's something special about doubling. If it's not apparent to you, doubling is what the 2s table is all about. Double 2 is 4, double 3 is 6, double 4 is 8, and so on through double 9, which is 18. By knowing your 2s table, you can figure out the 3s and the 4s. Because $2 + 1 = 3$ and $2 + 2 = 4$, $3 \times$ anything = 2 $\times$ a number + 1 $\times$ that number, and 4 $\times$ a number requires doubling twice.

So, $3 \times 8 = 2 \times 8 + 1 \times 8$.

That's $3 \times 8 = 16 + 8 = 24$.

By the same reasoning,

$4 \times 8 = 2 \times 8 + 2 \times 8$.

That's $4 \times 8 = 16 + 16 = 32$.

Alternatively, $4 \times 8 = 2 \times (2 \times 8) = 2 \times 16 = 32$.

Am I trying to tell you that you don't need to learn your multiplication facts in order to multiply? Not really, because if you know your multiplication facts you'll find it a lot easier to multiply, but with the knowledge of just a few tables and the understanding of what multiplication really is, you can multiply at least up to multiplication by 12 in just two steps. You're going to need to know just a little more than how to double. First, you're going to need to know how to multiply by 10, but I'll bet you already know how to do that.

Pop Quiz

1. Sebbie's 3 years old; Melissa is 10 times his age. How old is Melissa?

2. Jake has 10 chores to do; Alex has 4 times as many. How many chores does Alex have?

3. The average woman has 10 fingers on her hands. How many fingers do 5 average women have?

4. A basket contains 6 pounds of apples. How many pounds of apples are in 10 such baskets?

Answers

Hopefully, you answered the four questions above with the numbers 30, 40, 50, and 60, and having done so, you should recognize the pattern of the 10 × table. You just append 0 to the end of the number you're multiplying by 10.

Once you can multiply by 1, 2, and 10, in addition to being able to multiply by 3 and 4 in two steps, you can also multiply by 8, 9, 11, and 12 in two steps. Think about it.

$$8 \times = 10 \times - 2 \times$$
$$9 \times = 10 \times - 1 \times$$
$$11 \times = 10 \times + 1 \times$$
$$12 \times = 10 \times + 2 \times$$

(Yes, I am assuming that you know how to subtract one- and two-digit numbers. If I'm wrong, don't try it until after we've discussed subtraction.)

All we're missing now are the 5, 6, and 7 × tables. To get those, we're going to have to rely on one more that you probably already know.

That's the 5 × table. The 5 × table is one more (like doubling) that most of us pick up as kids and don't forget. That's because there's an easy-to-latch-onto pattern. Multiples of 5 alternately end with 5 (if an odd multiple) and 0 (if an even one). Also, we like to count by 5s early on: 5, 10, 15 . . . 50. If you're still shaky about your 5s, think "5 is half of 10." So if 10 × 8 is 80, 5 × 8 is half of that, or 40. Got it now? Once you have your 5 × on straight, think

$$6 \times = 5 \times + 1 \times$$
$$7 \times = 5 \times + 2 \times$$

So 5 × 7 = 35, 6 × 7 = 35 + 7 = 42, and 7 × 7 = 35 + 14 = 49.

Remember, these techniques don't take the place of learning your times tables, but they may help you to learn them, or give you something to fall back on. There's no way you can multiply larger numbers without having a grasp on those multiplication facts. That's why there are flash cards to cut out at the back of this book.

MULTIPLE-DIGIT MULTIPLICATION

Larger-number multiplication, like larger-number addition, can be done mentally from left to right. But it may sometimes require subtraction, so we'll leave it for later. With paper and pencil, multiplication follows the pattern below. To multiply 386 × 45, proceed as follows:

1. Line them up right justified.

$$\begin{array}{r} 386 \\ \times 45 \\ \hline \end{array}$$

2. Multiply 5 × 6.

$$\begin{array}{r} 386 \\ \times\ 45 \\ \hline \end{array}$$

3. Get 30, so write the 0 beneath the 5 and rename the 30 as 3 in the tens column:

$$\begin{array}{r} 3 \leftarrow 3 \text{ tens} \\ 386 \\ \times\ 45 \\ \hline 0 \leftarrow 0 \text{ ones} \end{array}$$

4. Next multiply 5 times 8 (really 5 times 80, but we pretend for simplicity's sake). Add the 3 in the tens column to get 43 tens. Write those 3 tens beneath the line, and rename the 40 tens into the hundreds place (up top).

$$400 \longrightarrow 4\,3$$
$$386$$
$$\times\ 45$$
$$\overline{30}$$

$$5 \times 8 \text{ tens} = 400$$

3 tens

5. The final multiplication by the 5 is times the 3 in the hundreds place for a total of 1500. To that we add the regrouped 400 to get a total of 1900 which we write below the line.

$$4\,3$$
$$386$$
$$\times\ 45$$
$$\overline{1930}$$

$$5 \times 300 = 1500$$
$$+ 400$$
$$\overline{1900}$$

6. Before we multiply by the 4, we're going to erase those "carry numbers" from our multiplications by 5. If you can't erase the ones you do on paper, cross them out. It will make it a lot less confusing for the new renamings you're going to have to do. Because we're next multiplying by 40, there will be no ones, so we place a "0" in the ones place on the next line of the answer. This is different, isn't it?

$$386$$
$$\times\ 45$$
$$\overline{1930}$$
$$0$$

Each multiplication done by the 4 (for simplicity's sake) is really being done by 40, and the placement of that 0 in the ones place serves to move everything over to the left by one column (that is, to make each number ten times what we're pretending it is).

7. Now multiply 4×6. You'll get 24 to distribute as shown.

$$2$$
$$386$$
$$\times\ 45$$
$$\overline{1930}$$
$$40$$

$$4 \text{ tens} \times 6 = 24 \text{ tens}$$

8. Moving along, the next multiplication will be 4 × 8, giving us 32, to which we need to add the renamed 2, like so:

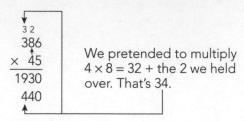

```
  3 2
  386
× 45
─────
 1930
  440
```

We pretended to multiply
4 × 8 = 32 + the 2 we held
over. That's 34.

9. Our last multiplication for this one will be 4 × 3. That's 12, to which we add the 3 that was waiting for us. Write that sum below the 19 and beyond to complete the multiplication portion of the solution.

```
   3 2
   386
 ×  45
 ─────
  1930
 15440
```

Finally 4 × 3 = 12 + 3 = ⑮

10. Of course, we're not yet done. I did mention that multiplication was related to addition, you'll recall. To cement that relationship, draw an addition line.

```
  386
× 45
─────
 1930
15440
─────
```
← Draw addition line . . .

11. Add to get the sum that is really the product (what we call an answer to a multiplication).

```
    386
 ×  45
 ─────
  1930
 15440
 ─────
 17,370
```
 . . . and add.

I've also added a comma for your reading pleasure. That's 17 thousand, 3 hundred seventy.

Whew! Are you still with me? The odds (we'll speak about odds later) are that you're never going to need to do a multiplication like that one while standing on one foot or when you are without a calculator. Of course, if you stick a calculator into your pocket, you'll never be without one. Why not try a few multiple-digit multiplication questions while you're still calculatorless, just for the heck of it? And be sure to think before you solve.

Pop Quiz

1. Myles donated $345 to charity 29 separate times. How much did he donate in all?

2. Kira bought 47 souvenirs for each of her 886 closest friends. How many souvenirs did she buy in all?

3. A total of 693 people spent $57 each on circus tickets. How much money was collected by the ticket seller?

4. The baseball bleachers were filled to capacity with 784 spectators 62 times this year. If a hot dog vendor had sold a hot dog to each of the spectators at all those games, how many hot dogs would he have sold?

Answers
Here are the solutions worked out.

1.
$$\begin{array}{r} 345 \\ \times\ 29 \\ \hline 3105 \\ 6900 \\ \hline 10{,}005 \end{array}$$

Hence, he donated $10,005.

2.
$$\begin{array}{r} 886 \\ \times\ 47 \\ \hline 6202 \\ 35440 \\ \hline 41{,}642 \end{array}$$

3.   693
     × 57
     ‾‾‾‾
     4851
    34650
    ‾‾‾‾‾
    39,501

Therefore, $39,501 was collected.

4.   784
     × 62
     ‾‾‾‾
     1568
    47040
    ‾‾‾‾‾
    48,608

That's a lot of hot dogs.

Hey, Hey, Take It Away

Subtraction (also traditionally known as **taking away**) is the first of the uncombining, or taking apart, operations. Subtraction undoes addition, but there is really no reason why anybody should ever have to subtract. Subtraction can be performed as addition, as used to be the case at all grocery stores before the advent of computerized cash registers that tell the cashier how much change to give. Say you bought something that cost $3.85, and you handed the cashier a $10.00 bill. The cashier would hand you a nickel and a dime followed by a $1 and a $5 bill and say, "$3.85 + .15 = $4.00 + $1.00 makes $5 + $5.00 makes $10." I call this **subtraction by making change.** To find out the difference between what the item cost and what your change is, add up all the amounts the cashier counted out for you.

$0.15
1.00
5.00
‾‾‾‾‾
$6.15

That means the difference between $10.00 and $3.85 is $6.15. That also could have been written as

$10.00
− 3.85
‾‾‾‾‾
$ 6.15

That's right, it's a subtraction, but we never took anything away. (By the way, difference and remainder are two different names for the answer to a subtraction.) You never need to be able to take away in order to subtract in columns either. Just keep in mind that the subtraction facts are the same as the addition facts, only sort of inside-out. One addition fact that you already know (I hope) is 5 + 3 = 8. The equivalent subtraction fact would be 8 − 5 = 3 or 8 − 3 = 5. But you don't need to approach a subtraction as something foreign. When facing 8 − 5 = what(?), think of it as 5 + what? = 8. You already know that 5 + 3 = 8, so 3's the answer. Yes, it's that simple.

Standard Column Subtraction

You actually can see a subtraction in column form by looking just above the preceding paragraph. That look should also let you in on the fact that (just like column addition and multiplication), it's right justified. Consider finding the difference between 875 and 432. It's customary to write the larger number above the smaller one, so that difference would be written like this:

$$875$$
$$-432$$

Starting in the right-hand column, ask yourself either "5 take away 2 leaves what?" or "2 + what? makes 5." In either case, the answer is 3, so we write that below the line beneath the 2.

$$87\mathbf{5}$$
$$-43\mathbf{2}$$
$$\overline{\mathbf{3}}$$

Next, move over to the tens column. There you can ask yourself one of four questions. Treating the numbers in the tens place as if they were completely isolated, you could ask

1. "7 take away 3 leaves what?"
2. "3 + what makes 7?"

Or, considering the digits' place values,

3. "70 take away 30 leaves what?"
4. "30 + what makes 70?"

Regardless of whether your answer is 4 or 40, you're going to write a 4 below the line in the tens place.

$$875$$
$$-432$$
$$\underline{4}3$$

Finally, shift your attention to the hundreds place. However you do it, the difference between 8 and 4 is 4. That gets written below the 4.

$$\mathbf{8}75$$
$$-\mathbf{4}32$$
$$\underline{\mathbf{4}}43$$

So the difference between 875 and 432 = 443. To check the answer to this and any other subtraction, add up the bottom two numbers, and if you're correct you'll get the top number.

Adding up from right to left, in the ones, 3 + 2 = 5, in the tens, 4 + 3 = 7, and in the hundreds, 4 + 4 = 8. Pretty cool how that subtraction worked out so neatly, isn't it? Don't get too cocky, it rarely will.

Renaming in Subtraction

Though it is customary to place the larger numeral on top in a column subtraction, that is no assurance that each digit in the top number will be larger than the digit beneath it. For example, 817 is considerably larger than 348, but when they're aligned for column subtraction, a strange thing occurs:

$$817$$
$$\underline{348}$$

One can't take 8 from 7 and get a meaningful result (because we don't yet know about negative numbers). There's nothing we know can be added to 8 to make 7. The same is also true of the combination of digits in the tens place. Clearly this calls for some drastic action—revolutionary even! That action is known as **renaming** or **regrouping**. You may have learned it as "borrowing," but forget that. That bit of mumbo jumbo is single-handedly the most responsible term for making math incomprehensible to people, with the possible exception of misrepresenting fractions.

Because we can't take 8 from 7 (play along), we're going to go to the tens place in 817, take that one 10 that's there, remove it, and regroup it with the seven ones to make seventeen ones, like so:

$$
\begin{array}{r}
8\,0\,{}^{1}7 \\
-3\,4\,8 \\
\hline
\end{array}
$$

Now we can subtract in the ones column by asking ourselves what's the difference between 17 and 8, or, alternatively, 8 + what makes 17. Either way, we get

$$
\begin{array}{r}
8\,0\,{}^{1}7 \\
-3\,4\,8 \\
\hline
9
\end{array}
$$

I'm sure you've noticed that the situation in the tens column has gotten even more woeful than it was before—or has it? Before we couldn't subtract 4 from 1, now we can't subtract 4 from 0. That's not a major difference, especially when we have all those hundreds sitting right next door. Let's use them.

$$
\begin{array}{r}
{}^{7}\!\!\not{8}\,{}^{1}0\,{}^{1}7 \\
-3\,4\,8 \\
\hline
9
\end{array}
$$

We'll take one of those hundreds from the eight hundreds on top and rename it as ten tens (which is the same as one hundred, right?). Notice that the top row of this subtraction is still worth 817, even though it doesn't look like it. That's because seven hundreds and ten tens and seventeen ones . . . well, you add them up. Convinced yet? They still total 817. It just looks kind of weird.

Now there's nothing to keep us from subtracting in the tens place.

$$
\begin{array}{r}
{}^{7}\!\!\not{8}\,{}^{1}0\,{}^{1}7 \\
-3\,4\,8 \\
\hline
6\,9
\end{array}
$$

Find the difference between 10 and 4, or 4 + what makes 10? Whichever way you look at it, there's going to be a 6 under that 4. Finally, finish it off with the hundreds.

$$
\begin{array}{r}
{}^{7}\!\!\not{8}\,{}^{1}0\,{}^{1}7 \\
-3\,4\,8 \\
\hline
4\,6\,9
\end{array}
$$

What + 3 = 7, or 7 take away 3 is what? Yep, it's a 4, just like you thought it was. But let's not leave this problem just yet. Check the sub-traction by adding up each column from right to left. So 9 + 8 = 17; pretend we wrote the 7 (which is already there) and regrouped the ten into the next column to add to the 6. Now 1 + 6 = 7 and 7 + 4 = 11. Imagine writing the 1 atop the tens column and renaming the ten tens as one hundred below the 4. Now, 1 + 4 = 5; 5 + 3 = 8, so we have 817, and that's what we started out with. Ta da!

Have you seen this subtraction for long enough? I want to see it one more time.

$$817$$
$$-\,348$$

Suppose we had approached this as a subtraction by adding up; you know, the making-change model. Say we're buying something that costs $348 dollars and for some inexplicable reason we hand the salesman $817. Now, in making change, the salesman says, "Here's 2 to make 350, and 50 to make 400. Then 417 makes $817." Add up those amounts.

2 + 50 = 52; 52 + 417 = 469, and that's the answer!

Pop Quiz

Solve these by adding up.

1. You pay the counter person $140 for an item that costs $67. How much change would you get?

2. Jim collected 673 comic books, but decided to sell 486 of them. How many comic books did he have left?

Solve these using renaming.

3. Out of 823 spectators at a ballpark, 468 of them ate pizza. How many spectators did not eat pizza?

4. A hardware store carries two brands of paint. Out of the 900 gal-lons sold one week, 594 were Brand A. How many gallons of Brand B were sold?

Answers

1. $73

   67 + 3 = 70

   70 + 70 = 140

   3 + 70 = 73

2. 187

486 + 4 = 490

490 + 10 = 500

500 + 173 = 673

4 + 10 + 173 = 187

3. 355

4. 306

The solution to Question 3 was solved in the same way as the previous model problem by renaming or by adding up. Problem 4 can be easily solved by adding up, but the renaming solution needs some explanation.

$$
\begin{array}{r}
9\,0\,0 \\
-5\,9\,4 \\
\end{array}
$$

Because there are no tens to be able to rename, we need to go straight to the 900.

$$
\begin{array}{r}
{}^{8}\!\not{9}\,{}^{1}0\,0 \\
-5\,9\,4 \\
\end{array}
$$

Rename one of those hundreds as ten tens. That leaves us with eight hundreds and ten tens, which is just another way to write 900. Next, we go to those ten tens to get one of them for the ones place:

$$
\begin{array}{r}
{}^{8}\!\not{9}\,{}^{9}\!\not{1}\!\not{0}\,{}^{1}0 \\
-5\,9\,4 \\
\end{array}
$$

Now we have eight hundreds, nine tens, and ten ones. Is that still 900? You bet it is, so subtract.

$$
\begin{array}{r}
{}^{8}\!\not{9}\,{}^{9}\!\not{1}\!\not{0}\,{}^{1}0 \\
-5\,9\,4 \\
\hline
3\,0\,6 \\
\end{array}
$$

As shown, 10 – 4 = 6, 9 – 9 =0, and 8 – 5 = 3, for a difference of 306. Add up, just to check that out.

A House Divided . . .

The second uncombining operation is **division,** which is the most misunderstood whole-number operation. Some view division's relationship to subtraction the same as multiplication's relationship to addition. It is subtracting the same number over and over again until the starting number has been "used up," or caught up to. Others view division as the undoing of multiplication, since dividing one number by another gives you the number that what you're dividing by was multiplied by to get the number being divided. Does that sound confusing? Well, it is. Before trying to sort it out, though, the question that must be dealt with is which of those two views is correct. The answer, of course, is that they both are. I know that's just what you were hoping to hear.

Let's get some terminology straight, because that'll make it much easier to understand what we're talking about. There are two general ways to represent a division. Either way, any division has 3 parts.

The number being divided by is called the **divisor.**

$$120 \div 6 = 20 \qquad 6\overline{)120}^{\,20}$$
$$\underset{\text{divisor}}{\underbrace{\qquad}}$$

The number being divided into is called the **dividend.**

$$120 \div 6 = 20 \qquad 6\overline{)120}^{\,20}$$
$$\underset{\text{dividend}}{\underbrace{\qquad\qquad}}$$

The answer to a division is called the **quotient.**

$$\overset{\text{quotient}}{120 \div 6 = 20} \qquad 6\overline{)120}^{\,20}$$

We can look at a division as an undoing of multiplication, as in $40 \div 8 = \underline{\quad}$ means 40 divided by 8 equals what, or 8 times what? equals 40. We can also treat it as if it asks "How many 8s can be subtracted from 40?" Any one of those questions properly solved will give a quotient of 5. You probably recognize that $8 \times 5 = 40$, but did you ever stop to think about this?

$$40 - 8 = 32; \; 32 - 8 = 24; \; 24 - 8 = 16; \; 16 - 8 = 8; \; 8 - 8 = 0$$
$$\quad\; 1 \qquad\qquad 2 \qquad\qquad 3 \qquad\qquad 4 \qquad\qquad 5$$

Count 'em!

The fact of the matter is the $35 \div 5 =$ __ format is one that you'll probably never see, except as an academic exercise. It's usually used only with numbers the size of the division facts, (less than 144), which, not coincidentally, are the same as the multiplication facts. You are much more likely to use the bracket, a.k.a. "the long division bracket."

It is known as "the long division bracket" not because of its length, but because it is used to exercise the process known as "long division." But I'm getting ahead of myself. Let's first contemplate "short division."

SHORT DIVISION

Short division is a peculiar process, really an introduction to the process of long division, yet often given too short shrift. In actuality, it has its uses. Specifically, it can be used where a dividend (of any number of digits) is being divided by a single-digit divisor. Consider this:

$$9\overline{)91863}$$

The first question here is 9 goes into 9 how many times, or 9 times what? makes 9. The answer is 1, so write a 1 in the quotient above the dividend's leftmost digit.

$$9\overline{)\overset{1}{91863}}$$

The next question to ask yourself is how many 9s there are in 1. Well, you know that there aren't any, so the next digit that's going to be placed above that 1 is 0.

$$9\overline{)\overset{10}{91^1863}}$$

Because the 1 has not yet been divided, it is grouped with the next figure, to make 18. Take a moment to digest that.

Okay, you can also look at how the little 1 is written in front of the 8 to emphasize the fact that it's 18 we're next going to divide. So 18 divided by 9 is how much (or 9 times what?) makes 18.

$$9\overline{)\overset{10\,2}{91^1863}}$$

It's not too difficult to recognize that there are two (2) 9s in 18, with nothing left over, so we'll stick a 2 into the quotient above the 18 in the dividend.

Now the question becomes how many times 9 goes into 6. Hopefully, you realize that the answer is no times, and we represent that by placing a 0 above the 6.

$$\begin{array}{r} 10\ 20 \\ 9\overline{)91^{1}86^{6}3} \end{array}$$

The 6 has not yet been counted in the total, so it is now grouped together with the 3 in order to make the number 63. Think about that for a second. When you're done with that mental exercise, ask yourself how many 9s there are in 63, or what times 9 = 63. Hopefully you came out with the same answer that I did.

$$\begin{array}{r} 10\ 20\ 7 \\ 9\overline{)91^{1}86^{6}3} \end{array}$$

That's right, 91,863 divided by 9 is 10,207, or 10,207 × 9 = 91,863, which, in case you haven't figured it out yet, is a very good way to check your answer. Check subtraction by adding; check division by multiplying.

As previously noted, most divisions are not going to come out perfectly. Try doing this one. Don't worry; I'll wait.

$$273 \div 8 = \underline{}$$

In order to solve this division, it should be rewritten with a bracket.

$$8\overline{)273}$$

Asking how many 8s there are in 2 yields the result 0, but we never write 0 as the leftmost digit of a number, so we'll move on and group the 2 with the 7 that follows it. How many 8s are there in 27?

$$\begin{array}{r} 3 \\ 8\overline{)27^{3}3} \end{array}$$

Three 8s make 24, which is as close to 27 as we're going to get, so we'll put a 3 in the quotient above the ones digit of the number we were dividing into (yes, I know, it's a convenient fiction, but we did treat the 7 as the ones digit of 27). Because 3 of the 27 remain undivided (27 − 24 = 3), we group those 3 with the next digit in the dividend to make 33.

Now, ask yourself how many 8s there are in 33, (or 8 × what?) gets you as close to 33 as possible without going over (just like *The Price is Right* TV show). The closest we can get is 32, and it takes four 8s to get us there.

$$\begin{array}{r} 3\ 4 \\ 8\overline{)27^{3}3} \end{array}$$

Okay, you say, but what about the 1 that remains when we've covered the four 8s (33 − 32 = 1). I'm glad you asked. There is, in fact, 1 remaining to be dealt with. It is known as a **remainder.** If you recall, "remainder" is the answer in a subtraction; "quotient" is the answer in a division. Very true, but how did we come up with that 1?

$$8\overline{)27\,^33}\;\;\genfrac{}{}{0pt}{}{3\;4\,\mathbf{r1}}{}$$

Sometimes, you'll see the "r" capitalized. Either way, it stands for remainder and was arrived at by subtraction. If you happen to be a purist and don't want to mix subtraction into your division, feel free to divide the 1 by the divisor, 8. Then you'll get a pure quotient.

$$8\overline{)27\,^33}\;\;\genfrac{}{}{0pt}{}{3\;4\,\frac{1}{8}}{}$$

That's right, placing the remainder over the divisor gives us a fraction to be tacked onto the end of the quotient and complete the division. So how do you know when you need to use a fraction or when you should use a remainder, or does it not matter? Well there are two answers to the last question: Sometimes it matters, sometimes it doesn't. If you're dealing with pure numbers, it doesn't matter. On the other hand, if you're dealing with real problems, the problem itself will suggest the way to display the answer.

For instance, let's say you have nine feet of wire and you wish to divide it into four pieces of equal length. How long would each part be? Well, there are four 2s in 9. Would it make sense to say each piece will be two feet long and we'll have a one-foot remainder? I don't think so. That's four pieces of equal length and a fifth piece of a different length. This doesn't satisfy the criterion stated by the problem, namely to cut the nine feet of wire into four pieces of equal length. How about each piece will be 2¼ feet long? This solution satisfies the criterion.

Now suppose you were going to have nine guests at a wedding and you want to spread them equally among four different tables at the reception. How many guests will you put at each table? How about 2¼? Can you possibly cut one of those guests into four portions (and before you say something cute, this is not Lizzie Borden's wedding reception)? The answer is of course not. So what's the alternative? Use the division answer, 2r1, that is two guests at each of the four tables, and one guest remaining to be seated somewhere else. You might argue that doesn't solve the problem either, but it's as close to a solution as there is.

Pop Quiz

Find the quotients. Express each answer in the form that makes the most sense.

1. At Talia and John's wedding, 140 guests are to be seated at 9 round tables in the banquet hall. How many will sit at each table?

2. You have 123 pounds of cheese to be cut up into eight packages of equal weight. What will each package weigh?

3. Ninety-four automobiles are to be distributed by the manufacturer equally among seven dealerships. How many automobiles will each dealership receive?

Answers

1. 15r5
2. 15⅜ lbs.
3. 13r3

Note that in Problem 1, the answer had to be in remainder form. 5 guests will have to be squeezed in, or another table found. For the second problem, each package had to weigh the same, so a fractional expression was required. In Problem 3, the car might be worth more if chopped up for parts, but dealerships and manufacturers don't do that, so a remainder was inevitable.

THE LADDER

Ladder division combines division's relationships with both multiplication and subtraction, and is a rather informal way of dividing with multi-digit divisors. Consider the following:

$$37\overline{)854}$$

First estimate what the quotient should be. That's so when you do work out the solution, whether with pencil and paper or calculator, you'll know whether or not that solution is in the ballpark. Estimating for division is not really difficult to do. If you're dividing something in the two thousands by something in the two hundreds you should expect a quotient around 10. If you're dividing a number in the ten thousands by a number in the hundreds, your quotient should be in the hundreds, since 100 hundreds make 10,000. Always estimate before you divide and then see how your quotient stacks up against that estimate.

Here, we're dividing 800 something by almost 40. How many 40s are in 800? Too hard a question? Then lop the last 0 off of both of them and ask yourself how many 4s there are in 80. It's really the same question in a slightly different form. Did you get a quotient of 20 and a little more? That's close enough for an estimate. Keep it in the back of your mind, or pencil it into the margin while we work the problem.

This is not something you'd want to try solving by short division, but you don't have to get too formal either. Look at the dividend and the divisor: 854 divided by 37. Well, you know there are at least ten 37s in there, so let's write them under the 854 and subtract.

$$37 \overline{)854} \\ \underline{370}|10 \\ 484|$$

The 10 out on the side keeps track of how many 37s you're subtracting. 854 − 370 = 484. You could call 484 the remaining dividend, and 10 the partial quotient.

Next, since I know there are another ten 37s in 484, I'll subtract another ten 37s:

$$37 \overline{)854} \\ \underline{370}|10 \\ 484| \\ \underline{370}|10 \\ 114|$$

We have a new partial quotient out on the step of "the ladder," and the remaining dividend is 114.

How many 37s do you think there are in 114? I'm going to go out on a very safe limb and guess 3.

$$37 \overline{)854} \\ \underline{370}|10 \\ 484| \\ \underline{370}|10 \\ 114| \\ \underline{111}| \; 3 \\ 3|$$

Pretty good guess. Heck, it was a great guess. Oh, so I multiplied in my head $3 \times 30 = 90$ and $3 \times 7 = 21$; $90 + 21 = 111$, before I selected 3. You caught me. The remainder is 3, and for the quotient we find the sum of the numbers on the rungs of the ladder.

$$
\begin{array}{r|l}
37 \overline{)854} & \\
370 & 10 \\
\hline
484 & \\
370 & 10 \\
\hline
114 & \\
111 & 3 \\
\hline
R3 & 23 \\
\end{array}
$$

How does that quotient compare to the estimate we made before solving the division? Pretty close, eh?

Don't think that you needed to do it the way it was done above. The beauty of the ladder form of division is that you could pull out two or five groups of 37 at a time. It'd just be a much taller ladder by the time you'd finished. Or, if you'd felt comfortable enough to pull out twenty 37s at a time, it could have been a shorter ladder, like this:

$$
\begin{array}{r|l}
37 \overline{)854} & \\
740 & 20 \\
\hline
114 & \\
111 & 3 \\
\hline
R3 & 23 \\
\end{array}
$$

Here's one more, with even bigger numbers.

$$293 \overline{)9345}$$

First, estimate the quotient. Go ahead, I'll wait. Done? Basically, you should have reasoned 9345 rounds down to 9000, and 293 rounds up to 300. So how many 300s are there in 9000? Beats me, so I'll knock two 0s off of each and ask myself how many 3s there are in 90. (Knocking two 0s off of both [dividing both by 100] leaves the relationship unchanged, a feature that is really exploited when solving algebraic equations, but I digress.) Because there are three 3s in 9, there are thirty 3s in 90. Also, because the divisor has been rounded up and the dividend rounded down, there are going to be a few more than thirty 293s in 9345, two or three more in all likelihood, for a quotient of 32 or 33. Now we're ready to divide.

I have sufficient conviction in my estimate to go straight to pulling thirty 293s out. I multiplied 293 by 3, and stuck a zero on to the end of it, like so:

$$293 \overline{)9345}$$
$$\underline{8790} \,|\, 30$$
$$555 \,|$$

Subtracting leaves 555 as the partial dividend, and I don't know about you, but it's pretty obvious to me that there are not going to be two 293s in that.

$$293 \overline{)9345}$$
$$\underline{8790} \,|\, 30$$
$$555 \,|$$
$$\underline{293} \,|\, 1$$

Finally, we subtract to get the remainder and add the partial quotients to get the quotient, 31.

$$293 \overline{)9345}$$
$$\underline{8790} \,|\, 30$$
$$555 \,|$$
$$\underline{293} \,|\, 1$$
$$R262 \,|\, 31$$

You could express your quotient as 31r262 or $31^{262}/_{293}$, which is pretty darned close to our estimate of 32.

Pop Quiz

Find the following quotients using ladder division. Estimate each before you solve. Express each remainder in the most appropriate form.

1. A 582 square foot deck is 24 feet wide. How long is it?

2. 8782 ounces of rice are divided equally among 372 sacks. How many ounces of rice does each sack contain?

3. 12,853 drills are to be divided up for shipping among 431 pallets. How many drills will go on each pallet?

Answers

1. 24¼ ft.
2. 23^{113}⁄₁₈₆ oz.
3. 29r.84

The fractions in answers to 1 and 2 have been simplified. One possible solution for each is shown below:

1.
```
24 )582
    480|20
    ───
    102|
     96|  4
    ──────
     R6|24
```

2.
```
372 )8782
    7440|20
    ────
    1342|
    1116|  3
    ──────
   R226|23
```

3.
```
431 )12853
    8620|20
    ────
    4233|
    2155|5
    ────
    2078|
    1724|4
    ──────
   R354|29
```

TRADITIONAL GUHZINTUHS

There is absolutely no reason except closure to take note of the "guhzintuhs," a term that teachers frequently use for traditional long division. The term came into favor as a result of the total confusion of students trying to learn the technique. They would attempt to say "this goes into that so many times," with numbers substituted for the "this," the "that," and the "so many times," but because they seemed to be very unsure of what they were doing, they would slur their word so that it sounded like "7 guhzintuh 28 four times," giving birth to the derisive term for long division.

There is nothing you can do with traditional long division that you can't accomplish with the ladder method, but, you probably had a horrible experience with it the first time around, so I deem it incumbent upon me to give you a second, albeit unnecessary, look.

Long division starts out looking like the divisions that we just solved by the ladder method. The solution, however, is quite different. Just for purposes of comparison, I'm going to use the same example used with the ladder.

$$37\,\overline{)854}$$

For openers, we're going to create a trial divisor by first rounding the real divisor to the nearer ten and then using only the tens digit, like so: 37 to the nearer ten is 40, so the trial divisor will be 4. Now we're going to divide that 4 into the first digit of the dividend, 8. It goes in twice, so we're going to use 2 as the first digit of the quotient. Because the divisor is a two-digit one, we'll place the partial quotient, 2, above the second digit of the dividend, like so:

$$\begin{array}{r} 2 \\ 37\,\overline{)854|} \end{array}$$

Are you still with me? If you're not, go back and read the description of what was done again. If you're with me now, hang on. We now multiply the partial quotient (2) times the divisor (37) to get 74, which we write under the leftmost two digits of the dividend.

$$\begin{array}{r} {}^{\otimes}\!\frown 2 \\ 37\,\overline{)854} \\ =\!\searrow\!74 \end{array}$$

Next, we'll subtract 74 from 85 to get 11.

$$\begin{array}{r} 2 \\ 37\,\overline{)854} \\ \underline{74} \\ 11 \end{array}$$

Hold onto your hat now, if you're wearing one. The next thing we do is bring down the 4 to make a new partial dividend of 114.

$$\begin{array}{r} 2 \\ 37\,\overline{)854} \\ \underline{74}\!\downarrow \\ 114 \end{array}$$

Now we're going to divide that trial divisor—remember the trial divisor, 4—into the first digit of the dividend, 1. The first digit of the dividend is smaller than the trial divisor, so we'll have to bundle that 1 with the second digit of the dividend to get 11.

We see that 4 goes into 11 two times, so I should use 2 as my next partial quotient—but a trial divisor doesn't give the correct partial product all of the time, and this happens to be one of those times. I'm savvy enough to know that 2 is too small a partial quotient, but suppose you didn't recognize that and used the 2. Then, you'd get the following situation:

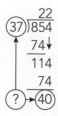

Multiplying the new partial quotient (2) times the actual divisor (37) will again give you 74 to write beneath the 114 with the columns aligned to the right. Draw a line and subtract and you'll get a remainder of 40! How can you have a remainder of 40 when the divisor is only 37? The answer is you can't. Red flag! Siren! Etc. Any time a remainder or partial dividend exceeds the size of the divisor it means that the partial quotient was not large enough. It's the signal to back up and reconsider the number above the 4 in the dividend.

$$
\begin{array}{r}
23 \\
37\overline{)854} \\
74\!\downarrow \\
\hline
114 \\
111 \\
\hline
3
\end{array}
$$

Ah, that's more like it! With a 3 over the 4 in the dividend, multiplying that 3 times the divisor yields 111, which, when subtracted from the 114, leaves a remainder of 3.

I'm not going to waste any more of your time or my time with this subject. Feel free to use this form of division if you like; I personally hope to never see it again. It's almost barbaric.

4

Common Fractions

You know what a common fraction looks like, and you may even know what it means. Most people only know part of what it means, and that's because it can mean so many things.

A common fraction has two parts: a top, called the **numerator,** and a bottom, called the **denominator.** They are separated by the fraction line—hey, that's a third part. No wonder fractions are so confusing. I tell you there are two parts and then throw in a third! Actually, there are only two parts you need to concern yourself with, and those are

$$\frac{\text{numerator}}{\text{denominator}}$$

We'll refer to common fractions (the kind with the two parts) simply as fractions from here on, but bear in mind that decimals and percents are also fractions—ones with special characteristics of their own.

What Is a Fraction?

A fraction may represent one equal part of a whole, as in $\frac{1}{2}$, $\frac{1}{4}$, etc., or more than one equal parts of a whole as in $\frac{2}{3}$, $\frac{3}{4}$, and so on. A fraction may also represent a ratio. If Henry has five acres of land and Karen has six acres of land then Henry and Karen have land in the ratio $\frac{5}{6}$. A fraction may also be a whole in itself, waiting to be subdivided. If you have one part of a pie that has been cut into eight equal pieces, then you have $\frac{1}{8}$ of pie, but you probably don't think of it as that. You're much more likely to think of it as a piece of pie. Mmm, that's a pretty yummy-looking piece of pie you have there. May I have half of it? Please?!

Here's the part of this introduction to fractions that I always hate. It's where I have to tell you that if your mother was anything like my mother, she probably lied to you about one very important thing. Did she ever cut a sandwich into two unequal parts and then tell you she was giving you the bigger half? Well, there is no such thing as a bigger half—at least not in *that* sense. Two halves of the same thing must be the same size. There is another sense possible, however, as you can see below:

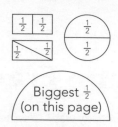

Biggest $\frac{1}{2}$
(on this page)

That being the case, we'll have to qualify the last paragraph and say halves of the same or congruent figures are equal in size. (**Congruent** is a fancy word meaning identical in shape and size.)

Fractions can represent many wholes as well, for example, $^4/_2 = 2$, $^9/_3 = 3$, $^8/_2 = 4$, and so forth. You see, every fraction is also a division example. Read the division line as "divided by" in each of the preceding fractions and you'll understand.

Not only is the realm of fractions (also known as rational numbers) infinite, but there are an infinite number of ways to express every fraction. How's that, you say? Well consider just the fraction $^1/_2$:

$^1/_2 = ^2/_4, ^3/_6, ^4/_8, ^5/_{10}, ^6/_{12}, ^7/_{14}, \ldots 1{,}000{,}000/2{,}000{,}000, \ldots$

And the same can be done for any and every fraction. Check out these.

Pop Quiz

1. Which of these is another way of representing $^1/_3$ of a pizza?

   $^2/_6$ $^3/_9$ $^6/_{18}$ $^8/_{24}$

2. Which of these is another way of writing $^1/_4$ pound of chopped beef?

   $^5/_{20}$ $^3/_{12}$ $^4/_{16}$ $^{25}/_{50}$

3. Which of these is another way of writing $^1/_5$ of a quart of milk?

   $^{10}/_{20}$ $^3/_{15}$ $^2/_{10}$ $^6/_{30}$

4. Which of these is another way of representing $^3/_4$ of your yearly income?

   $^9/_{12}$ $^7/_8$ $^{12}/_{16}$ $^{75}/_{100}$

Answers

1. all
2. all but $^{25}/_{50}$
3. all but $^{10}/_{20}$
4. all but $^{7}/_{8}$

Equivalent Fractions

As you just saw, there are many ways to write every fraction, and that situation demands some system to bring order to all that potential chaos. Fortunately, there is a way to do just that, and we'll get to it in a moment. Fractions that mean the same thing are called **equivalent fractions.** You saw many equivalent fractions in the pop quiz that ended the preceding section of this chapter. How could I ask you to work with equivalent fractions before introducing you to them? You were actually introduced to them in the last section; you just weren't told what they were called.

Any fraction can be changed to an equivalent fraction by multiplying its numerator and denominator by the same number. How can that be? Consider this:

$$\frac{3}{4} \times \frac{2}{2} = \frac{3 \times 2}{4 \times 2} = \frac{6}{8}$$

What's the meaning of $^{2}/_{2}$? Think of it as a division, $2 \div 2 = 1$. So if $^{2}/_{2} = 1$, then we just multiplied $^{3}/_{4}$ by 1, and as we both know, multiplying something by 1 does not change its value; therefore, $^{6}/_{8} = ^{3}/_{4}$.

Now, surely it has occurred to you that if we can multiply both **terms** of a fraction by the same number, we ought to be able to divide both terms by the same number without changing the fraction's value. ("Terms" is a generic name for both numerator and denominator.) If it hadn't occurred to you, consider it occurred now. Dividing by 1 doesn't change things any more than multiplying by 1, but there is a distinct advantage. If the numerator and denominator of a fraction contain a common factor, it would be possible to express that fraction in simpler terms by dividing out that common factor. Consider $^{25}/_{100}$. Do you recognize a number that's common to both terms? How about 5?

$$\frac{25}{100} \div \frac{5}{5} = \frac{25 \div 5}{100 \div 5} = \frac{5}{20}$$

Hmm, $^5/_{20}$? I see another common factor there, and it's 5 again, so let's get rid of it.

$$\frac{5}{20} \div \frac{5}{5} = \frac{5 \div 5}{20 \div 5} = \frac{1}{4}$$

Now that's better. There's no way to express that fraction in terms simpler than those. But what if we'd recognized the 25 in there as a common factor in the first place? I know, you probably did, but play along with me.

$$\frac{25}{100} \div \frac{25}{25} = \frac{25 \div 25}{100 \div 25} = \frac{1}{4}$$

That was certainly more economical than the two separate steps we used before. There's a name for the role 25 plays in the fraction $^{25}/_{100}$. It's called the **Greatest Common Factor,** or **GCF.** Finding the GCF and removing it permits us to express a fraction in its simplest terms (also called lowest terms).

In the bad old days, people referred to expressing fractions in lowest terms as "reducing fractions," another sure-to-confuse jargon. We are not reducing the fraction; we are reducing the terms in which that same fractional relationship is expressed. The greatest common factor will not always be one of the terms of the fraction. Consider the fraction $^{36}/_{48}$.

The factors of 36 are: 1, 2, 3, 4, 6, 9, 12, 18, 36.

The factors of 48 are: 1, 2, 3, 4, 6, 8, 12, 16, 24, 48.

The factors common to both terms are clear to see, but what is the GCF? Obviously it's 12, so that's the number we'll use to simplify $^{36}/_{48}$.

$$\frac{36}{48} \div \frac{12}{12} = \frac{36 \div 12}{48 \div 12} = \frac{3}{4}$$

What do you know? It's $^3/_4$. Please, whatever you do, do not leave this lesson thinking that you have to use the GCF to simplify fractions. You could have divided $^{36}/_{48}$ by $^2/_2$ and gotten $^{18}/_{24}$, then divided $^{18}/_{24}$ by $^2/_2$ to get $^9/_{12}$, then divided $^9/_{12}$ by $^3/_3$ to get $^3/_4$, or done it in any other sequence or combination of common factors. The most important thing is to end up with the number in its simplest form. How you get there can be your little secret, but by using the GCF, you'll get there the most efficient—read that fastest—way.

Pop Quiz

Find the GCF for each of the following fractions. Then express each fraction in lowest terms.

1. $^6/_{32}$ of a gallon of milk
2. $^{20}/_{50}$ of a box of nails
3. $^{32}/_{48}$ of a crate of eggs
4. $^{34}/_{51}$ of all the United States and the District of Columbia

Answers

1. 2   In lowest terms, $^3/_{16}$
2. 10   In lowest terms, $^2/_5$
3. 16   In lowest terms, $^2/_3$
4. 17   In lowest terms, $^2/_3$

Adding and Subtracting Fractions

Adding and subtracting fractions is easy. Just look.

$^1/_4 + ^2/_4 = ^3/_4$
$^4/_5 - ^1/_5 = ^3/_5$
$^3/_{12} + ^8/_{12} = ^{11}/_{12}$
$^{15}/_{16} - ^5/_{16} = ^{10}/_{16} = ^5/_8$

Do you see the pattern? Add or subtract the numerators. The denominators stay the same. After all, the denominators name what it is you're adding or subtracting: fourths, fifths, twelfths, and sixteenths. The numerators tell how many of them you're adding or subtracting. Check out the following figures.

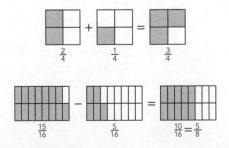

It looks pretty easy, doesn't it? Of course, there's one small detail that we haven't taken into consideration just yet. What do you do when you come upon something like this?

$$\left(\tfrac{1}{2}\right) + \left(\tfrac{1}{3}\right) = \,?$$

How should we treat $\tfrac{1}{2} + \tfrac{1}{3}$?

In the first examples we saw, each pair of fractions being added or subtracted had the same denominators. These two fractions do not. You already know how to write equivalent fractions for any other fraction, and that fractions with the same denominator can be added or subtracted, so you already know how to solve this addition. You just need to put the two pieces of knowledge together. What we need here are equivalent fractions for one-half and one-third, which have the same denominator. Any ideas?

We can find a common denominator for any two fractions by multiplying their denominators together. With our denominators, $3 \times 2 = 6$, so sixths is a common denominator for both thirds and halves.

$$\left(\tfrac{1}{2}\right) = \left(\tfrac{1}{6} \,\tfrac{1}{6}\right) \quad \tfrac{3}{6}$$

$$\left(\tfrac{1}{3}\right) = \left(\tfrac{1}{6} \,\tfrac{1}{6}\right) \quad \tfrac{2}{6}$$

To change halves to sixths, first divide the new denominator by the old.

$6 \div 2 = 3$

Then multiply the old numerator by that quotient to get the new one.

$1 \times 3 = 3$

Therefore:

$$\frac{1}{2} = \frac{3}{6}$$

Doing the same thing for one-third

$6 \div 3 = 2$

$1 \times 2 = 2$, so

$$\frac{1}{3} = \frac{2}{6}$$

Finally, add the new fractions:

$$\frac{3}{6} + \frac{2}{6} = \frac{5}{6}$$

That may seem like a lot of work, but if you followed the process, it was all very logical.

1. Find a common denominator.
2. Divide the new denominator by the old.
3. Multiply the quotient from (2) by the old numerator.
4. Add or subtract the new fractions.

But wait—it's not soup yet.

I told you that a common denominator can be found by multiplying the denominators of the fractions you're trying to add or subtract together, and that's quite true, but . . . The "but" is that you can end up with a denominator that's quite large, awkward, and much greater than the one that's necessary. For example, suppose you needed to subtract $\frac{3}{8}$ from $\frac{29}{32}$. The common denominator you'd arrive at by multiplying the two together is 256ths, but, in actuality, 32nds would be just fine. Because 32 is divisible by 8, just expressing the $\frac{3}{8}$ as $\frac{12}{32}$ would have done the trick. I got that by dividing 32 by 8, getting the quotient 4, and multiplying the numerator 3 by that, which gave me 12. So

$$\frac{29}{32} - \frac{12}{32} = \frac{17}{32}$$

Thirty-two is not just **a** common denominator for 8ths and 32nds, but it is **the least common denominator,** or **LCD.** By using the LCD when writing equivalent fractions, you assure that the fractions you are working with are in lowest terms (read that easiest to work with).

Pop Quiz

Find the LCD for each of the following pairs of fractions. Then solve and express each answer in lowest terms.

1. What is $\frac{3}{8}$ teaspoon of salt – $\frac{1}{3}$ teaspoon of salt?
2. How much is $\frac{1}{6}$ cup of butter + $\frac{3}{4}$ cup of butter?
3. Combining $\frac{1}{9}$ pound of flour + $\frac{1}{4}$ pounds of flour yields _____?
4. Taking $\frac{5}{12}$ pounds plaster of Paris from $\frac{15}{16}$ pounds plaster of Paris leaves _____?

Answers

1. 24ths; $\frac{1}{24}$ teaspoon of salt
2. 12ths; $\frac{11}{12}$ cup of butter
3. 36ths; $\frac{13}{36}$ pound of flour
4. 48ths; $\frac{25}{48}$ pounds plaster of Paris

There is yet another way to find least common denominators. That method is known as **prime factor trees,** and I'm not going to go into that here. That's because almost every fraction you encounter is likely to be 32nds or larger, so that you'll never have a need for prime factor trees. If you'd like to see how they work anyway, look in Chapter 17 of my *CliffsNotes Parents Crash Course Elementary School Math* (Wiley Publishing, Inc., 2005).

Multiplying Fractions

Multiplication of fractions is strange in only one way. Before justifying that observation, however, let me say that it is the easiest operation to learn. That's because it follows the straight rules for multiplication of whole numbers—twice. Here's an example:

$$\frac{2}{3} \times \frac{5}{8} = \frac{2 \times 5}{3 \times 8} = \frac{10}{24} = \frac{5}{12}$$

Notice that the numerators are multiplied together and the denominators are multiplied together. Finally, the product is expressed in lowest terms.

Now here comes the strange part: The product, $5/12$, is less than each of the numbers being multiplied together, so multiplying fractions, unlike multiplying whole numbers, is not a combining operation. To more fully understand what multiplying fractions is, let's substitute the word "of" for the × sign; after all, 2 of 2 is 4 and 2 of 3 is 6, etc.

$$\frac{1}{2} \text{ of } \frac{1}{2} = \frac{1}{4}$$

Multiplying fractions is taking a part of a part. Half of one-half is one-fourth. Does that make a bit more sense now?

There's one more twist to multiplying fractions, which can make things even easier. The traditional term for the operation is **canceling.** To show how canceling works, consider the first fractional multiplication. When we got to the end, we had to simplify the answer to express it in lowest possible terms. Here's the same multiplication using canceling.

$$\frac{2}{3} \times \frac{5}{8} = ?$$

$$\frac{\overset{1}{\cancel{2}}}{3} \times \frac{5}{\underset{4}{\cancel{8}}} = \frac{5}{12}$$

Before multiplying we look for common factors in the numerators and the denominators. In this case, 2 is common to both 2 and 8, so we factor (divide) it out of both: $2 \times 1 = 2$, $2 \times 4 = 8$. Finally, we multiplied 1×5 and 3×4 to get $5/12$. The simplification was done before multiplying instead of afterward. Here's one more of those.

$$\frac{9}{16} \times \frac{12}{27} = ?$$

$$\frac{\overset{1}{\cancel{9}}}{\underset{4}{\cancel{16}}} \times \frac{\overset{\cancel{3}^{1}}{\cancel{12}}}{\underset{\underset{1}{\cancel{3}}}{\cancel{27}}} = \frac{1}{4}$$

Because 9 and 27 are both divisible by 9, we cancelled them first. Next, 12 and 16 are divisible by 4. But that leaves $3/3$ on the right side of the × sign. They cancel to make 1. Finally, $1 \times 1 = 1$ and $1 \times 4 = 4$. Try a few on your own.

Pop Quiz

Multiply each of the following pairs of fractions. Express each answer in lowest terms. (Bear in mind that the word "of" is synonymous with the instruction to multiply.)

1. Find $^3/_8$ of $^1/_3$ cup of shortening.
2. Kira wants to make $^5/_6$ of a recipe that calls for $^3/_{20}$ pound of sugar. How much sugar should she use?
3. Emile needs $^4/_9$ of $^9/_{12}$ dozen eggs. How many dozen eggs does he need?
4. How long is $^{25}/_{32}$ of $^{16}/_{30}$ inch?

Answers

1. $^1/_8$ cup of shortening
2. $^1/_8$ pound of sugar
3. $^1/_3$ dozen eggs
4. $^5/_{12}$ inch

Reciprocals

Believe it or not, there's a good reason for studying reciprocals at this juncture. An understanding of reciprocals is necessary to understand division of common fractions. The reciprocal of a number is what you must multiply it by to get a product of 1. Not getting it? Then consider this. The reciprocal of $^1/_2$ is 2. That's because $2 \times ^1/_2 = 1$. The reciprocal of $^1/_3$ is 3; the reciprocal of 3 is $^1/_3$. The reciprocal of $^3/_5$ is $^5/_3$; the reciprocal of 100 is $^1/_{100}$; $^5/_8$'s reciprocal is $^8/_5$.

Are you thinking to yourself, "Self, to get the reciprocal of a number that's a fraction, all I have to do is turn it upside down, and for a whole number, I put a one over it." If that is what you're thinking, then you've got it down pat—or whatever your name is. Try these.

Pop Quiz

Find the reciprocal of each of the following. Don't bother to write the answer in lowest terms (if applicable).

1. $\frac{3}{8}$
2. $\frac{5}{6}$
3. 17
4. 49

Answers

1. $\frac{8}{3}$
2. $\frac{6}{5}$
3. $\frac{1}{17}$
4. $\frac{1}{49}$

Dividing Fractions

There is no ambiguity about whether division of fractions is more closely related to multiplication or subtraction. It is the undoing operation of multiplication. Dividing a fraction by another fraction will yield a quotient that is larger than the number being divided into.

The mechanics of fractional division rely on your knowing which is the **divisor** and which is the **dividend.** Because the ÷ sign means is divided by, in a division written in the form "(1st number) ÷ (2nd number) = ?", the second number is the divisor.

Now comes the good part, and why we need reciprocals. In order to divide one fraction by another when written in the preceding form rewrite the division as a multiplication by the reciprocal of the divisor. "Huh?" you ask. Consider the following:

$$6 \div 2 = 3$$

Change that to a multiplication by the reciprocal of the divisor and get

$$6 \times \frac{1}{2} = 3$$

See? It works for whole numbers, although we rarely use that fact, but we could if it were the most efficient way to divide them. For fractions, it's the only way. Let's try one.

$$\frac{3}{8} \div \frac{2}{5} = ?$$

$$\frac{3}{8} \times \frac{5}{2} = \frac{15}{16}$$

If you were tempted to cancel the 2 and the 8, don't do it. Canceling may be used only on numerator and denominator of the same fraction or across a × sign.

Check out this one:

$$\frac{3}{16} \div \frac{5}{8} = ?$$

$$\frac{3}{\cancel{16}_2} \times \frac{\cancel{8}^1}{5} = \frac{3}{10}$$

Do you understand why that canceling was legal? It was done across the "×" sign.

Here are a few for you to try your hand at.

Pop Quiz

Solve each of the following divisions. Make sure each quotient is in lowest terms.

1. Alex needs to divide ⅛ cup of sugar by ⅓. How much will he get?
2. Tziona needs to divide ⅚ tablespoon of vanilla extract by ¹⁰⁄₁₂. What will be the result?
3. Juanita divides ⁴⁄₉ of her working hours by ¾, resulting in _____?
4. Sebastian must divide ²⁵⁄₃₂ cord of wood among ¹⁵⁄₁₆ of his customers. How much will each get?

Answers

1. ⅜ cup
2. 1 tablespoon
3. ¹⁶⁄₂₇
4. ⅚ cord

Fractions Greater than One

While solving the fractional divisions in the last Pop Quiz, you encountered fractions greater than one—in fact, you created fractions greater than one when you wrote the reciprocals of the divisors. Any fraction with a numerator equal to its denominator has a value of one ($\frac{2}{2}$, $\frac{3}{3}$ . . . , $\frac{100}{100}$, etc.). Any fraction with a numerator greater than its denominator has a value greater than one: $\frac{3}{1}$, $\frac{12}{10}$, $\frac{4}{3}$, and $\frac{16}{15}$ are the four such fractions used in the aforementioned quiz. Such a fraction would usually be expressed as whatever whole numbers it contains, and the fractional remainder, so the four preceding fractions would be (in simplest form) 3, $1\frac{1}{5}$, $1\frac{1}{3}$, and $1\frac{1}{15}$. With the exception of the 3, which is a whole number, those are known as **mixed numbers.** A mixed number is part whole number, and part common fraction.

The simplest way to create a mixed number from a fraction greater than one is to divide the numerator by the denominator, and write the remainder as a fraction:

$$\frac{23}{4} = 5\frac{3}{4}$$
$$\frac{31}{7} = 4\frac{3}{7}$$
$$\frac{5}{2} = 2\frac{1}{2}, \text{ etc.}$$

Adding Mixed Numbers

To add mixed numbers, simply add the whole numbers and add the fractions. If the fractions have unlike denominators, rewrite them as equivalent fractions with a common denominator. Should the fractions add up to a fraction greater than one, rename it as a whole number, and add the whole number part to the whole number part of the sum. Here's an example:

$$
\begin{array}{c|c}
5\frac{3}{4} & 5\frac{6}{8} \\
2\frac{3}{8} & 2\frac{3}{8} \\
\hline
7\frac{9}{8} & 8\frac{1}{8}
\end{array}
$$

The least common denominator for 4ths and 8ths is 8ths, so we change the $\frac{3}{4}$ to $\frac{6}{8}$. Because the lower fraction is already in 8ths, we just rewrite it into the next column to align for addition. Next, we add to get $7\frac{9}{8}$, but $\frac{9}{8}$ simplifies to $1\frac{1}{8}$, which we then add to the 7 to get $8\frac{1}{8}$.

Here's another one.

$$\begin{array}{c|c} 2\frac{3}{10} & 2\frac{9}{30} \\ 4\frac{13}{15} & 4\frac{26}{30} \\ \hline & 6\frac{\cancel{35}}{30}\ \ 7\frac{1}{6} \end{array}$$

To add $2\frac{3}{10} + 4\frac{13}{15}$ we need to find the LCD for 10ths and 15ths. That turns out to be 30ths. So $\frac{3}{10}$ is rewritten as $\frac{9}{30}$; $\frac{13}{15}$ becomes $\frac{26}{30}$. Adding, we get $6\frac{35}{30}$. That's the same as $6 + 1\frac{5}{30}$, which simplifies to a total of $7\frac{1}{6}$. Try a few.

Pop Quiz

Solve each of the following additions. Make sure each sum is in simplest terms.

1.

$$\begin{array}{c|} 3\frac{3}{4} \\ 2\frac{1}{2} \\ \hline \end{array}$$

2.

$$\begin{array}{c|} 2\frac{1}{3} \\ 3\frac{1}{4} \\ 2\frac{5}{6} \\ \hline \end{array}$$

3.

$$\begin{array}{c|} 3\frac{5}{12} \\ 2\frac{3}{6} \\ 3\frac{5}{8} \\ \hline \end{array}$$

Answers

1.

$$\begin{array}{c|c} 3\frac{3}{4} & 3\frac{3}{4} \\ 2\frac{1}{2} & 2\frac{2}{4} \\ \hline & 5\frac{\cancel{5}}{4}\ \ 6\frac{1}{4} \end{array}$$

2.

$$\begin{array}{c|c} 2\frac{1}{3} & 2\frac{4}{12} \\ 3\frac{1}{4} & 3\frac{3}{12} \\ 2\frac{5}{6} & 2\frac{10}{12} \\ \hline & 7\frac{\cancel{17}}{12}\ \ 8\frac{5}{12} \end{array}$$

3.

$$\begin{array}{c|c} 3\frac{5}{12} & 3\frac{10}{24} \\ 2\frac{3}{6} & 2\frac{12}{24} \\ 3\frac{5}{8} & 3\frac{15}{24} \\ \hline & 8\frac{\cancel{37}}{24}\ \ 9\frac{13}{24} \end{array}$$

Subtracting Mixed Numbers

Everything that was true about adding mixed numbers applies to subtracting mixed numbers, with two exceptions. First, there is no limit to the number of mixed numbers that may be added together, as you just got reminded in the last quiz, but subtraction is limited to two numbers. Second, it may be necessary to rename in order to subtract.

The first example of subtraction is as straightforward as it can get.

$$7\frac{8}{9} \qquad 7\frac{8}{9}$$
$$-5\frac{3}{9} \qquad -5\frac{3}{9}$$
$$\overline{} \qquad \overline{2\frac{5}{9}}$$

We simply subtract the lower fraction from the upper one and subtract the lower whole number from the upper one and there's the difference. I'm not going to bother you with examples where the fractions have to be changed because the denominators are different. They work exactly the same way as in additions, except, of course, you subtract. But what about something like this?

$$8\frac{3}{7}$$
$$-5\frac{6}{7}$$

There is no way to subtract $^6/_7$ from $^3/_7$. There just aren't enough 7ths in the top number. But we're going to do something about that. First we'll rename one whole from the 8 as 7ths, thusly:

$$8\frac{3}{7} \;\Big|\; 7\frac{7}{7}$$
$$-5\frac{6}{7} \;\Big|$$

Of course there's still a matter of the $^3/_7$ that we started out with. That gets added to the $^7/_7$ to make $^{10}/_7$.

$$8\frac{3}{7} \quad\bigg|\, 7\frac{7}{7} \quad\bigg|\, 7\frac{10}{7}$$
$$-5\frac{6}{7} \quad\bigg|\quad\quad\, 5\frac{6}{7}$$
$$\overline{\phantom{-5\frac{6}{7}}\quad\bigg|\quad\bigg|\, 2\frac{4}{7}}$$

Now we can subtract, and get the difference, 2⁴⁄₇.

I'll do one more for you, with all the bells and whistles that it's possible to throw in.

$$9\frac{1}{3}$$
$$-4\frac{3}{4}$$

Clearly, we can neither add nor subtract 3rds and 4ths, so we'll have to find a common denominator. The LCD is 12ths. So ¹⁄₃ becomes ⁴⁄₁₂; ³⁄₄ becomes ⁹⁄₁₂.

$$9\frac{1}{3} \quad\bigg|\, 9\frac{4}{12}$$
$$-4\frac{3}{4} \quad\bigg|\, 4\frac{9}{12}$$

Now we can see that ⁹⁄₁₂ cannot be subtracted from ⁴⁄₁₂, so we'll have to rename the 9 as 8¹²⁄₁₂; then we'll add the ¹²⁄₁₂ to the ⁴⁄₁₂ that were already there.

$$9\frac{1}{3} \quad\bigg|\, 9\frac{4}{12} \quad\bigg|\, 8\frac{16}{12}$$
$$-4\frac{3}{4} \quad\bigg|\, 4\frac{9}{12} \quad\bigg|\, 4\frac{9}{12}$$
$$\overline{\phantom{-4\frac{3}{4}}\quad\bigg|\quad\bigg|\, 4\frac{7}{12}}$$

Finally, we subtract the 4 from the 8 and the ⁹⁄₁₂ from the ¹⁶⁄₁₂ to get 4⁷⁄₁₂.

Pop Quiz

Solve each of the following subtractions. Make sure each difference is in lowest terms.

1. From the $9\frac{7}{8}$ sheets of plywood Myles had yesterday, he sold $6\frac{5}{8}$. How many does he have left?

2. Tania had $8\frac{1}{4}$ skeins of wool at the beginning of the week, but used $6\frac{5}{6}$ of them. How many remain?

3. From $10\frac{2}{5}$ pounds of flour $5\frac{7}{10}$ were used to bake. How much is left?

4. Friday morning Jakob had $12\frac{5}{8}$ studs, but he used $8\frac{15}{16}$ to frame a dividing wall. How many studs remained?

Answers

1. $3\frac{1}{4}$ sheets

2. $1\frac{5}{12}$ skeins

3. $4\frac{7}{10}$ pounds

4. $3\frac{11}{16}$ studs

Changing Mixed Numbers to Fractions

Earlier in this chapter we discussed changing fractions greater than one to mixed numbers (really mixed numerals, but let's not get technical) by dividing the numerator by the denominator and writing the remainder as a fraction. Sometimes it's desirable to change a mixed number to a fraction larger than one.

To change a mixed number to a fraction, first multiply the whole number part by the denominator of the fraction; then add the numerator and place the result over the denominator, as in the diagram below.

$$\otimes \;\; 2\frac{3}{8} \;\; \overset{①}{②} \; \xrightarrow{} \; 16 + 3 \; \xrightarrow{} \; \overset{③}{} \; \frac{19}{8}$$

Let's try another one. To change $5^4/_7$ to a fraction greater than one, first multiply the denominator times the whole number; that's $7 \times 5 = 35$. Add the product to the numerator; $35 + 4 = 39$, and put the result over the denominator to get $^{39}/_7$.

"Why would anyone want to change a mixed number to a fraction?" you might well ask. The answer to that question is in the next section, once you get a little practice.

Pop Quiz

Express each mixed number as a fraction greater than one.

1. $3^4/_5$ cups of coffee
2. $4^5/_8$ hours
3. $5^4/_9$ pounds of ham
4. $6^8/_{15}$ annoying math problems

Answers

1. $^{19}/_5$ cups of coffee
2. $^{37}/_8$ hours
3. $^{49}/_9$ pounds of ham
4. $^{98}/_{15}$ annoying math problems

Multiplying and Dividing Mixed Numbers

This section is going to be much shorter than you probably anticipated when you saw its title. The fact of the matter is there's no way to multiply or divide mixed numbers. Don't read that as saying mixed numbers can't be multiplied or divided. They just can't be *while they're mixed numbers*. And that's what the preceding section was all about. In order to divide or multiply mixed numbers, first change them to fractions greater than one. Then follow the rules for multiplying and dividing fractions. If the product or quotient turns out to be larger than one, express it as a mixed number. In any case, always express the answer in lowest terms. Here are a few to practice on.

Pop Quiz

Solve each of the following problems. Make sure each product or quotient is in lowest terms.

1. $3\frac{3}{8} \times 4\frac{2}{3} = ?$
2. $5\frac{5}{6} \times 6\frac{7}{12} = ?$
3. $7\frac{4}{9} \div 2\frac{3}{4} = ?$
4. $9\frac{5}{8} \div 3\frac{15}{16} = ?$

Answers

1. $15\frac{3}{4}$
2. $38\frac{29}{72}$
3. $2\frac{70}{99}$
4. $2\frac{4}{9}$

5

Decimal Fractions

You use **decimal fractions** every day, since our system of dollars and cents is based on them. "Deci-" is a prefix meaning tenth, and so decimal fractions extend the place value system of numeration to the right, where a decimal point separates the whole numbers from the fractions. The whole numbers are to the left of the decimal point and the fractions to the right. But wait—I may be assuming too much.

Place Value

If you know all there is to know about **place value,** feel free to skip this section. Otherwise, here it is in a nutshell—through all the whole numbers you're ever likely to need.

It is place value that permits us to represent numbers of infinite or infinitesimal size using just ten digits. Using those ten digits, we put them into various places or imaginary columns named **H, T,** and **U** (the H, T, and U stand for hundreds, tens, and units (ones) respectively). The worth of the digit is its name times the column heading times the **period** in which that digit appears. What's a period? Starting to the left of the decimal point and working left from there, the periods are ones, thousands, millions, billions, trillions. . . . There are more periods left of trillions, but you're never going to need them. Even the government doesn't—yet. If you look at the figure below, you can see that the imaginary lines, column headings, and period names have been drawn in.

	TRILLIONS			BILLIONS			MILLIONS			THOUSANDS			ONES			
	H	T	U	H	T	U	H	T	U	H	T	U	H	T	U	Periods
(1)												4		6	5	
(2)										3		7		2	1	
(3)								4	6		2	9			7	
(4)				2	7	1				4	1	6		9	1	
(5)	5	6	7	4	9	6	1	6	9	2		4	3		6	
(6)	9												7			
(7)		3	5							1				6	7	

Place-value chart

The number on line (1) is read four thousand sixty-five. Notice that the name of the period is not used for the ones period, since it has no bearing on the value of the number. (Multiplying by one causes no change in value.) Also notice that the name of the T place is usually expressed as a "ty" tacked onto the number in the column, except for twenty and thirty (which have their own unique names). Forty isn't spelled "fourty," and fifty isn't spelled "fivety," but you can tell where they came from readily enough. If there were no columns it would look like this: 4065. The zero is used as a **placeholder.** It takes the place of the column with nothing written in it once the number has begun being read. You begin reading the number from its leftmost digit. Placeholders are not used to the left of the leftmost digit. It used to be customary to use a comma to separate the periods, like this: 4,065. Some books still represent it that way. It is more in vogue, however, to not use commas until the number contains at least five digits, as you'll see in line (2) and after.

The number on line (2) is read three hundred seven thousand, twenty-one. Written in place-value form it would read 307,021. Place-holders were needed for the ten thousand and hundreds places.

Line (3)'s number is forty-six million, twenty-nine thousand, seven. It would be written in place-value form as 46,029,007.

Pop Quiz

Read the numbers on lines (4) through (7) by beginning with the leftmost digit, and saying the appropriate column names followed by the period name (except for ones). Then write each number in place-value form.

Answers

4. Two hundred seventy-one billion, four million, sixteen thousand, ninety-one (271,004,016,091)

5. five hundred sixty-seven trillion, four hundred ninety-six billion, one hundred sixty-nine million, two hundred four thousand, three hundred six (567,496,169,204,306)

6. nine hundred trillion, seven hundred (900,000,000,000,700)

7. thirty-five trillion, one hundred thousand, sixty-seven (35,000,000,100,067)

Extending Place Value

Just in case you hadn't noticed, go back and look at the place-value chart. You should notice that as you move to the left, each column heading is worth ten times the value of the column heading to its immediate right. This pattern continues for as far as you wish to go. More importantly, for our current mission, the converse is true. That is to say, as you move from left to right, each column is worth one-tenth of the column to its immediate left. The pattern continues to the ones column, which is one-tenth the column to its immediate left, the tens—but the pattern does not end there.

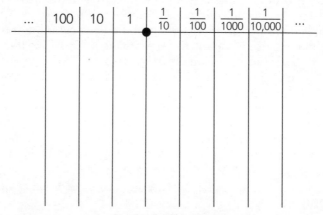

Decimal point chart

The decimal point serves to separate whole numbers from fractions, but the pattern continues across that decimal point. The first place to the right of the point is one-tenth the place to its immediate left, the next column to the right is one-tenth that, and so on without end, as indicated in the decimal point chart by the ellipses of three dots.

Note that decimal fractions are all expressed as multiples of tenths, such as tenths, hundredths, thousandths, etc. You could easily carry decimal fractions out to millionths, billionths, or trillionths, but except in the worlds of nuclear physics and engineering, there is no practical use for those fractions.

So far we've been all talk with no examples, so . . .

$$0.1 \qquad 0.02 \qquad 0.32 \qquad 0.670 \qquad 0.003 \qquad .402$$

Above, in order, are one-tenth, two hundredths, thirty-two hundredths, six hundred seventy thousandths (which is the same as 67 hundredths—more about that in a moment), three thousandths, and four hundred two thousandths. Notice that all but the last were written with a zero to the left of the decimal point. That zero is customary, but gratuitous. It just calls attention to the decimal point.

Now, more about that 0.670. Just as that 0 to the left of the decimal point is unnecessary, so is the 0 to the right of .67. Any zeroes to the right of the last significant digit (1 through 9) are gratuitous and not normally written, just as zeroes are not written to the left of the last significant figure in a whole number.

$$.67 = .670 = .6700 = .67000 = 0.67$$

They all mean the same thing: sixty-seven hundredths. Decimals that have the same values as mixed numbers are common, and they do not have any special names (such as mixed numbers vs. fractions). For example, 8.5 is eight and one-half, 6.25 is six and one-fourth, and 5.4 is is five and two-fifths. You may have to think about that last one, but $^4/_{10}$ simplifies to $^2/_5$.

Adding and Subtracting Decimals

I'm sure you're accustomed to adding and subtracting dollars and cents, and I'd like to be able to tell you that adding decimals is the same thing. There are similarities, and there are differences (no pun intended). If all decimals were two-place ones, there would be no difference.

$$
\begin{array}{r}
0.38 \\
+.49 \\
\hline
0.87
\end{array}
\qquad
\begin{array}{r}
0.63 \\
-.38 \\
\hline
0.25
\end{array}
$$

But as not all decimals confine themselves to two places, the key to adding and subtracting them is lining up the decimal points one above the other. So to add 0.9 + 1.33 + 2.513 we would write it as

$$
\begin{array}{l}
0.9 \\
1.33 \\
\underline{2.513}
\end{array}
$$

With the decimal points lined up in a single column, the digits fall so they are aligned in their proper places. Then, we add in column addition format.

$$
\begin{array}{l}
\overset{1}{0.9} \\
1.33 \\
\underline{2.513} \\
4.743
\end{array}
$$

The 3 in the thousandths column is all by itself, and so gets added to nothing else. There are a 1 and a 3 in the hundredths column that add to 4. In the tenths column, 9 + 3 + 5 sum to 17, so we write the 7 and rename the 10 tenths as 1 whole. Finally, after placing the decimal point in the answer we find that 1 + 1 + 2 = 4.

The key to all this is the lining up of the decimal points. Having done so, keeping your addition to only the numbers in the same place is also essential. If you find it difficult to keep your eye in the column when it is so ragged, you might want to add zeroes for help, like so:

$$
\begin{array}{l}
0.900 \\
1.330 \\
\underline{2.513}
\end{array}
$$

Those zeroes don't change the meaning of any of the numbers, but may make you less nervous about adding in columns.

Subtraction works pretty much the same way, except you can't have fewer or more than two numbers in a subtraction problem. Try finding the difference here.

$$\begin{array}{r} 3.25 \\ -1.8 \\ \hline \end{array}$$

What's the problem? You don't know how to subtract nothing from 5? Sure you do. Does this look better?

$$\begin{array}{r} 3.25 \\ -1.80 \\ \hline \end{array}$$

It means the same thing, so proceed.

1.

$$\begin{array}{r} 3.25 \\ -1.80 \\ \hline 5 \end{array}$$

2.

$$\begin{array}{r} {}^{2}\!\not{3}.{}^{1}25 \\ -1.80 \\ \hline 5 \end{array}$$

3.

$$\begin{array}{r} {}^{2}\!\not{3}.{}^{1}25 \\ -1.80 \\ \hline 1.45 \end{array}$$

1. Subtract 0 from 5 to get 5
2. Rename 3 wholes as 2 wholes and 10 tenths, which are regrouped to make 12 tenths
3. Subtract 8 tenths from 12 tenths to get 4 tenths. Place the decimal point and subtract 1 from 2 to get 1. The difference is 1.45.

Did you follow that? There's one other potential source of trouble when subtracting decimals. Can you figure out how to subtract 3.752 from 8.1?

$$\begin{array}{r} 8.1 \\ -3.752 \\ \hline \end{array} \longrightarrow \begin{array}{r} 8.100 \\ -3.752 \\ \hline \end{array}$$

The actual subtraction would then proceed with the appropriate renamings taking place:

$$\begin{array}{r} {}^{7}\!\not{8}.{}^{1}\!\not{0}{}^{9}\!\not{0}{}^{9}\!{}^{1}0 \\ -3.\,7\,5\,2 \\ \hline 4.\,2\,4\,8 \end{array}$$

Pop Quiz
Add or subtract as called for.

1. $3.5 million + $37.23 million + $0.359 million
2. 6.42 cords of wood – 4.897 cords of wood
3. 26.71 m of wire + 821.9 m of wire + 6.538 m of wire
4. 19.2 barrels of rum – 6.843 barrels of rum

Answers

1. $41.089 million
2. 1.523 cords of wood
3. 855.148 m of wire
4. 12.357 barrels of rum

Multiplying Decimals

For purposes of multiplying decimals, the decimal points are completely ignored until the very end. Consider the following: 1.3×0.48. First we'll stack the numbers for multiplying.

$$\begin{array}{r} 0.48 \\ \times 1.3 \\ \hline \end{array}$$

Next we'll multiply by the 3.

$$\begin{array}{r} 0.48 \\ \times 1.3 \\ \hline 144 \end{array}$$

Notice there's no decimal point in the partial product. We'll now go on to multiply by the one, first placing a zero as a placeholder in the rightmost column:

$$\begin{array}{r} 0.48 \\ \times 1.3 \\ \hline 144 \\ 480 \end{array}$$

Notice there is still no decimal point. Next, let's add those partial products:

$$0.48$$
$$\times 1.3$$
$$\overline{144}$$
$$\underline{480}$$
$$624$$

Now, it's finally time to get concerned about where that decimal point belongs. Count up the number of digits that are to the right of the decimal point in the factors being multiplied. There are two in 0.48 and one in 1.3. That's a total of three digits to the right of the decimal point in the numbers being multiplied. Well, that's the number of places there'll be to the right of the decimal point in the answer, so

$$0.48$$
$$\times 1.3$$
$$\overline{144}$$
$$\underline{480}$$
$$.624$$

If you prefer, you could have written that product as 0.624.

Just in case that answer doesn't make sense to you, consider what we were multiplying.

We tried to find 1.3 (which is $1\frac{3}{10}$) times 0.48.

Well, $1 \times 0.48 = 0.48$. Keep that in the back of your head.

$\frac{3}{10} \times 0.48 = \left(\frac{1}{10} + \frac{1}{10} + \frac{1}{10}\right) \times 0.48.$

But $\frac{1}{10} \times 0.48 = 0.048$.

That means that $0.048 + 0.048 + 0.048 = \frac{3}{10} \times 0.48.$

That's 0.144.

Finally, add that on to 0.48:

$$0.480$$
$$\underline{+.144}$$
$$0.624$$

The defense rests!

Pop Quiz

Solve these multiplications.

1. 3.5 box cars hold 37.23 tons of freight. How much freight is that in all?

2. There are 6.42 yards of ribbon on each of 4.3 spools. How much ribbon is that in all?

3. $0.71 is spent at the refreshment stand by 6.54 thousand people at the ballpark. How much money is spent in all?

4. Sam jogs at a speed of 0.042 miles per hour for 0.84 hours. How far does he jog in all?

Answers

1. 130.305 tons

2. 27.606 yds.

3. $4.6434 thousand or $4643.40

4. .03528 miles.

   Did you know to add that 0 before the decimal point in 4?

Dividing Decimals

If the rules governing dividing fractions were the same as those for multiplying them, I'd have put them into the last section, so that should tell you that they don't work the same way. But while the rules for dividing decimals are different from those for multiplying, they're just as easy and straightforward. The only rule, actually, is that you can't divide **by** a decimal. Note that the emphasis is on the word "by."

"So," you may justifiably ask, "what do I do about something like $0.6\overline{)3.66}$?"

First of all, bear in mind that it's the 0.6 that's the trouble-maker, not the 3.66. The solution is to change the divisor, 0.6, to a whole number by moving its decimal point one place to the right, to turn it into the whole number, 6.

"Whoa! You can't do that," I heard someone object. "That changes the relationship between the two numbers!"

That is absolutely true, but I can preserve the relationship by also moving the decimal point in the dividend (remember, the dividend is the number being divided up) one place to the right, like so:

$$0.6 \overline{)3.66} \text{ becomes } 6 \overline{)36.6}$$

Do you doubt that the relationship is the same? If so, consider this:

$$30 \div 5 = 6 \qquad \text{Multiply both by 10 to get } 300 \div 50 = 6$$

Moving the decimal point one place to the right, I'll remind you, is the same as multiplying both sides by 10. The relationship remains unchanged.

The next task in the solution of the division is to place the decimal point in the quotient. Here we return to addition and subtraction placement. That is, it goes directly above the dividend's decimal point.

$$6 \overline{)36.6} \qquad \text{Then we divide} \qquad 6 \overline{)36.6} = 6.1$$

Once the decimal point is placed in the quotient (remember, the quotient is the answer to a division problem), we see that short division will take care of this. 6 goes into 36 perfectly 6 times. Then 6 goes into 6 once. Notice that once the decimal point has been placed in the quotient, we don't pay attention to it during the actual process of dividing.

I'll show you one more example, using a somewhat more difficult set of numbers.

$$0.48 \overline{)1.296}$$

First, move the divisor's decimal point. (Now, you can't see it, but it's there!)

$$48 \overline{)1.296}$$

Then do the same for the dividend.

$$48 \overline{)129.6}$$

How did you know how many places to move it? You need two to make 0.48 a whole number; and what you do in the divisor, you must duplicate in the dividend. Next, place the decimal point in the quotient.

$$48 \overline{)129.\overset{\cdot}{6}}$$

It's time to divide. How many 48s in 129?

$$48 \overline{)\begin{array}{l} \overset{2.}{129.6} \\ -96 \end{array}}$$

Next, we multiply 48 × 2, place that under the 129, and subtract.

$$48 \overline{)\begin{array}{l} \overset{2.}{129.6} \\ -96 \\ \hline 33\ 6 \end{array}}$$

Having gotten 33, we next bring down the remaining 6 to make a new partial dividend of 336. How many 48s are there in 336? I'll guess 7.

$$48 \overline{)\begin{array}{l} \overset{2.7}{129.6} \\ -96 \\ \hline 33\ 6 \\ \underline{336} \end{array}}$$

The quotient is 2.7, or two and seven-tenths. (Or we could have used the ladder.)

Pop Quiz
Solve these divisions.

1. 30.1 miles are run in 3.5 hours. How many miles per hour is that?
2. Find 0.798 ounces of gold divided by 0.38.
3. 20.996 tons of steel are divided equally among 5.8 boats. What does each boat weigh?
4. 27.606 hm of electrical wiring was used during the construction of 4.3 identical houses. How much wire went into each house?

Answers

1. 8.6 mph
2. 2.1 oz.
3. 3.62 tons
4. 6.42 hm

6

Using Percents

Percents, decimals, and common fractions are three different ways of naming and writing fractions. The main difference is that while fractions and decimals are based on 1 naming one whole, percent is based on 100 naming one whole, so 1 = 100%, or the whole thing. Anything less than 100% is less than the whole thing. Something greater than 100% is more than the whole thing, so 200% is two whole things, 150% is one and a half. But while this is all very real, it is not very practical stuff.

In the real day-to-day world, there are four places where percents come into play. Those are taxes, tipping, sales (not to be confused with sales tax), and interest. Don't be bashful about using fractions to stand in for percents when it's convenient to do so. Remember, since 100% is one whole, 50% is one-half, and 25% is one-fourth.

Percent Strategies

Any fraction or decimal can be expressed as a percent, but that's an academic exercise. Practically, what you're going to want to do is to find a percent of any number. I could tell you how to do that, but you've heard it before and managed to forget it, so it makes more sense to give you a quick and easy way to find certain percents of any number. As I'm sure you'll remember, if you think about it, the purpose of a decimal point is to separate whole numbers from fractional parts of numbers—specifically tenths and hundredths. That means 35 is the same as 35.00, while 35.4 means thirty-five and four-tenths. So 35.6 means thirty-five and six-tenths, but 35.60 means thirty-five and sixty one-hundredths. I'll remind you that the last two numbers really mean the same thing, since sixty one-hundredths and six-tenths are equivalent fractions.

Here comes the good part. To multiply a number by 10, all one needs to do is to move its decimal point one place to the right, so $35.60 \times 10 = 356.0$. Check it on your calculator; I'll wait. To multiply a number by 100, all you have to do is to move its decimal point two places to the right, so $35.60 \times 100 = 3560$. You can go and check that out, too.

"What does this have to do with percents?" you're probably asking. Please, have a little patience. Sometimes I can't help thinking—and acting—like a teacher. Division and multiplication are opposites. That means if moving the decimal point to the right is the same as multiplication, then moving it to the left must be the same as division.

And so, in fact, it is: $27 \div 10 = 2.7$. Notice that even though there is no apparent decimal point in "27," since it is a whole number, the actual decimal point separating it from the decimal-fraction numbers is understood to be after the 7, even if it isn't actually written as in 27.0. Dividing by 10 moves that point one place to the left, resulting in 2.7. It also just so happens that dividing a number by 10 is the same thing as taking 10% of it, so 10% of 27 is 2.7. I think you're ready for a Pop Quiz.

Pop Quiz

1. What's 10% of $49?
2. Find 10% of $67.
3. How much of $346 is 10%?
4. $25 is 10% of what amount?

Answers

1. $4.90
2. $6.70
3. $34.60
4. $250. That's right. Question 4 was meant to trick you. Did it? In the first three questions you were asked to find 10% of a dollar amount, while in Question 4, you were given 10% of the dollar amount and asked to find the quantity it was 10% of, which can be done by moving the decimal point one place to the right. Hey, how easy do you want this to be? If I can't throw in a trick question from time to time you'll be lulled into complacency. We can't have any of that, can we?

Now what do you suppose is involved in finding 1% of a number? Put differently, if 10% of 49 is 4.9, what do you suppose 1% of 49 is? Did you figure it out yet? No need to rush. All right. The answer is 0.49 (the zero isn't really necessary, but it's a convention that I'll use throughout the book). To find one percent of a number, move its decimal point (whether shown or not) two places to the left. One percent of 235 is 2.35. See how the decimal point was moved two places to the left of where it would have been if it had initially been written? One percent of 69.3 is 0.693. One percent of 347.8 is 3.478. One percent of 3000 is 30. Do you get it? Let's find out.

Pop Quiz

1. What's 1% of $498?
2. Find 1% of $637.
3. 1% of 34.60 is how much?
4. $35 is 1% of what number?

Answers

1. $4.98
2. $6.37
3. $0.346 (Read that 34.6 cents.)
4. $3500. That's right, Question 4 was meant to trick you again. Did you fall for it this time? Since you find 1% of a quantity by moving the decimal point left two places, when you're given the 1%, you find what it's 1% of by moving the decimal point right two places. It's really not that hard a concept, but you do have to be on your toes. Since you've already come this far, you should find the next section to be of some interest.

Interest

There are basically two types of interest. There's the **interest that you pay,** such as on your mortgage, credit cards, and other loans (auto and the like), and there is the **interest that banks and bonds pay to you.** Typically, the first type is high and the second type is low. Within those two types of interest there are two other types of interest: simple and compound. **Simple interest** is called that because figuring it out is simple. A percent of interest is paid on a fixed amount of money at the rate of something per year. Say you deposit $1000 in the bank at a rate of 2% per year. Then the interest the bank would pay you would be determined by the formula

$$I = p \times r \times t$$

In that formula, I stands for interest, p is principal (the amount of money deposited), r is the interest *rate*, and t is time, usually expressed in years. Since the principal mentioned above is $1000, and the interest rate is 2% per year, after one year the principal would have accumulated twice as much as 1%, which you already know how to find. What's 1% of $1000? Now double that to find 2% of $1000. Because 1% of 1000 = 10, 1% of $1000 = $10. Double that, and you'll get $20. So, using this scheme, at the end of a year your bank account would contain $1020. Unfortunately, things seldom work that way.

It is far more likely that your bank will calculate **compound interest,** using a monthly or a quarterly basis. Were it to do so on a monthly basis, it would calculate it based on $1/12$ of 2% per month. Let's not go there. Rather, let's consider the quarterly basis. If your bank computes interest quarterly, then it figures $1/4$ of 2%, which is $1/2$ of 1% each quarter. Since $10 is 1% of your $1000 deposit, half of that, or $5 is $1/2$%. That means that after $1/4$ year the account will contain $1005. The deposited interest now also accrues interest, so the next quarter's interest will be computed on $1005. Figure 1% of $1005 is $10.05 and half of that rounded to the nearest penny is $5.03. We'll add that to our $1005 for a total of $1010.03 after two quarters of a year. Do you feel up to computing the total that will be in the account after two more quarters? If you do, skip the next paragraph and see whether you get the same result I do. If you don't you can always come back here and follow along.

After the next quarter, 1% of $1010.03 = $10.1003, half of which to the nearest penny is $5.05. Added to the previous $1010.03, that makes a new total of $1015.08. Finally, to complete the year, we take 1% of the new total and halve that. So 1% of 1015.08 is 10.1508. Half of that rounded to the nearest whole cent is 5.0754 = $5.08. Adding to our previous total, $1015.08 + $5.08 = $1020.16.

So by your bank's compounding the interest quarterly, you managed to turn your $1000 into $1020.16—a whopping 16 cents more than you got from simple interest. But whose pocket would you rather see that money in? Yours or the bank's? Also remember if that had been $10,000 in your account it would have been $1.60 more. If it had been $100,000 it would have been $16.00 more, and if you have $100,000 in the bank, 1) I'd really like to get to know you, and 2) Haven't you heard of annuities or mutual funds? Look into them.

> ## Calculating Compound Interest?
>
> The formulas for calculating compound interest are quite complex, and far beyond the scope of this book. I thought a handy, dandy tear-out card with some compound interest tables on them might do you some good, until I saw how complicated they are. To calculate compound interest on any amount, go online and enter "compound interest calculator" into a search engine. There are many tables available online for calculating all kinds of compound interest and loan amortizations.

Taxes

Among the most annoying and inescapable forces of governments is the price of their doing business. Whether for policing, picking up the trash, or running and building the educational system, it's an extremely taxing practice.

INCOME TAX

Most of us pay federal income taxes, some of us pay state income taxes, and some of us pay city income taxes as well. If you're not a math whiz—and you wouldn't be reading this book if you were—you'll use tax software or a tax preparer of some sort to prepare your federal income tax return. I'm pretty good at math, and yet I gave up preparing my own federal return about ten years ago. The rules are just too complicated. Frankly, tax preparation software costs somewhere between $35 and $50 at the time of this writing, and buying it and investing a day or so of your time answering the straightforward questions it asks you can save you hundreds of dollars over paying a professional tax preparer (who likely as not drives a truck eleven months out of the year).

SALES TAX

As of 2006, Alaska, Montana, New Hampshire, and Oregon do not charge sales tax on purchases of goods or services. The amounts charged by the other states range from as low as 4% (Hawaii) to as much as 9% (Louisiana and Tennessee). In some cases, these amounts include charges of some localities tacked on to what the state collects. Some states tax food only when it's purchased in a restaurant. Some states tax clothing while others do not. I'll assume that you know what is taxed in your state and locality, as well as what the tax rate is. Since it would be nice to have an idea of what you're going to have to pay at the cash register and not just what it says on the price tag (which rarely includes the tax), here's how to do it.

In a nutshell, what you pay is equal to the sum of the price tag and the tax.

$$Cost = price + tax$$

Sales tax is just one more application of the percentage process that we studied earlier in this chapter. Since the tax varies from 4 to 9%, as already noted, we'll look at each of those in turn, but first, consider the following. Do you really need to find the exact tax on an item with a price tag of $137.58? I'm not saying that you couldn't if you wanted to, but why bother with that $0.58? Figure the tax on $138 and know within a few cents what the final total will be. Again, everything revolves around moving that decimal point.

Suppose you bought that $138 trench coat at an outlet store in Tennessee, where the tax rate is 9%. First find 10% of $138 by moving the decimal point one place left ($13.80) and then 1% by moving it two places ($1.38).

Since 10 − 1 = 9, 10% − 1% = 9%.

That means 10% of $138 − 1% of $138 = 9% of $138.

$$\begin{array}{r} \$13.80 \\ \text{So subtract:} \quad -1.38 \\ \hline \$12.42 \end{array}$$

Finally, add the tax to the price to find out how much you'll really be paying:

$$\$137.58 + \$12.42 = \$150.00$$

Holy cow! I don't know what the odds were of its coming out to a round figure like that, but I'll bet they're not very good (and if you take my bet you're a sucker)! Actually, though, since you rounded up to estimate the tax, the actual figure will be a little lower.

Suppose you bought that coat where the sales tax on it was 5%. You could find the tax the hard way, or the easy way.

Hard way: 5% = 1% + 1% + 1% + 1% + 1%

So tax = $1.38 + $1.38 + $1.38 + $1.38 + $1.38 = $6.90

Easy way: $5\% = \dfrac{10\%}{2}$

So tax $= \dfrac{\$13.80}{2} = \$6.90$

Either way, the coat will cost $137.58 + $6.90 = $144.48

Pop Quiz

Calculate what the same trench coat would cost if the sales tax were 4%, 6%, 7%, and 8%. (Hint: Use what we've already found, and build on it.)

Answers

1. 4% = 5% – 1%, so $6.90 – $1.38 = $5.52
   Coat costs $137.58 + $5.52 = $143.10
2. 6% = 5% + 1%, so $6.90 + $1.38 = $8.28
   Coat costs $137.58 + $8.28 = $145.86
3. 7% = 5% + 1% + 1%, so $6.90 + $1.38 + $1.38 = $9.66
   Coat costs $137.58 + $9.66 = $147.24

   Note that we just found 2% of the coat's cost is $2.76, so:
4. 8% = 10% – 2%, so $13.80 – $2.76 = $11.04
   Coat costs $137.58 + $11.04 = $148.62

Tipping

Calculating a tip (known on Park Avenue or in Beverly Hills as a gratuity) is even easier than figuring sales tax. The amount to tip, however, varies in different situations. The one thing to always bear in mind when calculating the amount of a tip is the service that was performed for you. Was it done well and cheerfully? Then you might want to tip more than the customary amount. Was the service performed poorly, rudely, or grumpily? Then you might want to tip less, or in an extreme case, none at all.

TAXI CAB

Twenty percent is considered the usual appropriate tip for a taxi cab driver. How do you compute 20% of a taxi cab fare that comes to $19? Do you remember how to find 10% of any number? (Just move the decimal point left one place.) Well 10% of $19 is $1.90 (remember money always is written with two decimal places for the cents—unless it's written with none, as in $19, which doesn't make any cents—sorry about that). Obviously, 20% is twice as much as 10%, so double the $1.90 to $3.80. Finally, to compute how much to hand to the taxi driver, add that $3.80 to the $19, and get $22.80. Of course, you could also round that up to $23, which will get you a bigger "Thanks" from the driver. On the other hand, if the driver was rude to you, you might want to round it down to $22.

RESTAURANTS

Most persons do much of their tipping in restaurants (and much of their tippling in bars or at weddings). The rule of thumb at restaurants is that an average tip (for average service) should be about 15% of the check (which is really a misnamed bill) before tax. Most states (and some municipalities) add a sales tax, a restaurant tax, or a dining tax to the total of your food and drink consumed in a restaurant. You should know what percentage that is. You can use it as a shortcut to figuring out how much to tip. If service is particularly bad, I've been known to leave a restaurant without tipping. Wouldn't you know that the one time I did that I had to go back because I'd left my keys on the table?

Waitpersons (a.k.a. waitresses and waiters) make a substantial amount of their income from tips, so not leaving one should only be done if you were extremely dissatisfied with that person's performance. Never short the waitperson because you didn't care for the quality of the food. Let the restaurant manager or maitre d' know that. Also be aware that in Europe it is customary for the restaurant to include the gratuity in your bill, and you don't want to tip on top of your built-in tip.

For poor to very poor service, you might want to reduce the size of the tip to between 10 and 12% of the cost of the meal. For good to very good service, you might want to increase the size of your tip to between 18 and 20% of your total. Keep in mind that as the person leaving the tip, you are the one to decide how much the service was worth. If you're planning on returning to this restaurant, however, you're better off having a reputation for being an average to better-than-average tipper than a poor one. It's liable to get you better service next time.

T.I.P.S.

You may find this acronym useful: **T.I.P.S.**, in the service world, means "To Insure Proper Service."

The best advice I can give you in calculating the tip without thinking is to look at the tax you were charged. Is it a 6% tax? Then double it for a 12% tip, double it and add half again for a 15% tip, or triple it for an 18% tip. Was the restaurant tax 10%? Add half of it to itself for a 15% tip, or double it for a 20% one. Was the restaurant tax some other number like 8%? Play with it. Twice 8% is 16%. Don't know what percent the dining tax was, or worst of all, there was none? Then read on.

Start with the untaxed total of your meal and find 10% of it. By now, you should be able to move that decimal point in your head. The meal came to $42.19? Don't worry about the 19 cents. Round it down to $42. What's 10% of $42? Four dollars and twenty cents ($4.20). Now decide what percent you're going to tip. For 15% take $4.20 plus half again, or $6.30. For 20% double the $4.20 to $8.40. Want to tip more than 15%, but not 20%? With the knowledge that $6.30 is 15% and $8.40 is 20%, pick a number in between.

Suppose the meal came to $57.80 and you want to tip $17\frac{1}{2}$%. First, round that $57.80 up to $58. Then 10% of $58 is $5.80 and 5% is half that amount, or $2.90. So 15% is $5.80 + $2.90 = $8.70. 20% of $58 is twice $5.80, or $11.60. Halfway between 15% and 20% you'll find $17\frac{1}{2}$%, so the amount we're looking for is halfway between $8.70 and $11.60. Do you care if you get the exact amount? I sure don't. A dollar more than $8.70 is $9.70 and a dollar less than $11.60 is $10.60. So $17\frac{1}{2}$% is about $10. 'Nuf said!

Pop Quiz

1. If a jacket had a $125 price tag and the sales tax was 6%, how much would you have to pay to buy the jacket?

2. If a taxi meter reads $35.25 and you want to tip the driver the conventional amount, how much should you give her?

3. If a restaurant had a dining tax of 8% and your meal's tax came to $1.20, what's a fair tip?

4. At the diner, your family ran up a $45 tab. How much money should you leave on the table so as to include a 15% tip?

Answers

1. Ten percent of the price tag is $12.50, 5% is half of that, or $6.25, and 1% is $1.25. Add 5% and 1% to get the 6% sales tax: $6.25 + $1.25 = $7.50. Add that to the price of the jacket for a total of $132.50.

2. A conventional taxi cab tip is 20%. Round the amount down to $35 and find 10% of it: $3.50. Double that to make 20%: $3.50 + $3.50 = $7.00. Give the cabbie $42.25.

3. Just double the sales tax for a nice 16% tip of $2.40.

4. Figure 10% of $45 is $4.50. Find 5% by halving that, $2.25. So 15% is $4.50 + $2.25 = $6.75, which, when added to $45, makes $51.75.

On Sale, or Discounts

It may seem like semantics, but I'm sure you know that there's a big difference between "For Sale" and "On Sale." Just about everything in this country is for sale if the price is right. But our purpose here is to determine what we should expect to save (not to be confused with a savings account)—in other words, how big is the discount—and what we should expect to pay when something is **on sale.** Since we have spent so much time learning how to compute sales tax, we're going to disregard taxes in the next three sections.

WHAT YOU CAN SAVE

"Biggest Sale of the year," the ad trumpets. "Save 40% on everything in the store." Well what exactly does that mean? Suppose I buy an item that is normally $50. Since I'm going to get 40% off of that $50, I use the trick we've been using throughout this chapter (and will continue to use throughout the book).

First find 10% of $50 by moving the decimal point 1 place to the left:

10% of $50 = $5.00.

Now, I could go one of two ways.

1. If I'm looking for the amount off that 40% is, I simply multiply 10% by 4. After all, 40% is 4 × 10%:

4 × $5 = $20. So I can take $20 off the normal $50, meaning I pay $50 − $20 = $30.

Or, I could have taken my thinking in another direction:

2. If I'm taking 40% off the $50 item, I'm actually paying 60% of the normal price. Remember, 100% − 40% = 60%.

Don't get nervous. The more ways you have of looking at something, the more options you have for finding the solution easily. So, 60% of $50 is 6 × 10% of $50: 6 × $5 = $30. Well what do you know? I got the same answer both ways! I told you there was more than one way to do it.

Let's consider another item—this one lists for $100, but this time the sale is 35% off. Do you see a quick way of doing this, involving no math whatsoever—well, hardly any?

What's 10% of $100? Why, it's $10. Coincidence? I think not! What's 20% of $100? It's $20, and you can arrive at that answer any way you like. Do you see where I'm going with this? If you do, then think about this one: What's 35% of $100? I sure hope you said $35; in fact, any percent of $100 will be the same dollar amount as the percent amount! So the sale price is going to be 65% of $100 (100% − 35% = 65%), or $65.

Now that ought to suggest something to you when dealing with larger numbers—especially multiples of $100. Let's see if it does. What is 29% of $400? Think about it this way. If $29 is 29% of $100, then 4 times 29 is 29% of $400: 4 × $29 = $116. That means that a 29% discount on a $400 item would bring the price down to $400 − $116 = $284.

Finally, for this section, a pair of boots that usually sells for $119.95 has been discounted 39%. How much should you expect to pay for them? To solve this, we want nothing to do with $119.95, so round it to $120, and remember, if you're a few cents off, so what? You're using what you know to learn to find practical solutions. It's not as if you're going to be tested on it next period. Next, find 10% of $120 ($12) and 1% of $120 ($1.20). The amount we're looking for, 39%, is 40% − 1%. That's 4 times 10%, less 1%:

The parentheses show that we multiply first, then subtract:	39% = (4 × 10%) − 1%
Now let's put in the actual numbers (of means ×):	39% of $120 = (4 × $12) − $1.20
Next, multiply:	39% of $120 = ($48) − $1.20
And, finally, subtract:	39% of $120 = $46.80

But we're not quite through. You were asked how much you should expect to pay. To find that, you need to subtract that $46.80 from the list price of about $120 to get about $73.20. Can you figure out another way to have solved the same problem, using maybe a step or two fewer? If you're getting a 39% discount, you're paying 61% of the list price (100% − 39% = 61%).

$$61\% \text{ of } \$120 = 6 \times \$12 + \$1.20$$
$$\text{which} = \$72 + \$1.20$$

Hmm, looks like $73.20!

Pop Quiz

1. About how much money can you expect to save at a 60% off sale on an item listing for $29.95?
2. About how much would you pay for a pair of $90 shoes at 45% off?
3. If a $230 stereo is offered at a 25% discount, how much should you expect to save?
4. What would you spend if the savings is 37% off a $400 suit?

Answers

It is important to understand the question. Questions 1 and 3 ask how much you'll save, while questions 2 and 4 ask how much you'll spend (requiring an extra step if you first find the discount).

1. Figure out 60% of $30, which is 6 × $3, or $18.
2. You are paying 55% (100% – 45% = 55%). First, 50% of $90 is one half, or $45. To find 5%, move the decimal one more place ($4.50). Now add the two and get $49.50.
3. You could find 10%, 10%, and 5%, or you could recognize that 25% is the same as one-fourth. One-fourth is one-half of one-half, so take half of $230, which is $115, and half of $115 = $57.50.
4. We're dealing with hundreds of dollars. If you got 37% off a $100 suit, you'd pay $63. Remember, % means per hundred, and if you're saving 37 per hundred, you're paying 100 – 37 = 63. For $400, multiply 63 × 4 = twice 63 ($126) and twice that = $252. Don't be shy about improvising, as long as you're on solid ground.

DISCOUNTING THE DISCOUNT

Sometimes you'll see a store offering discounts on already discounted merchandise. "Take an additional 20% off merchandise already discounted 30%," the ad will proclaim. Well, an additional 20% off something already discounted 30% is like getting 50% off, isn't it? In fact, it is not! When something has been discounted 30%, it's selling for 70% of its usual price. Discounting it another 20% means marking its price down to 80% of the 70% it's now selling at. Let's say it's a $100 item currently selling for $70. Ten percent of $70 is $7, so we'll be reducing the price by 2 × 7, or $14, and $70 – $14 = $54. So reducing by 20% after it had originally been reduced by 30% is equivalent to discounting the original price by 46%. (Remember, 100% works just like $100, and $100 – $54 = $46.) Feel free to work with other combinations of double discounts using $100 as your base price, and you'll discover that the second discount never makes the entire reduction the same as if the store had taken the combined amounts off the original price.

WHAT WAS IT ORIGINALLY?

This is more of an academic exercise than the practical ones that we've dealt with before, but sometime or other you just might actually want to know how much a discounted item cost originally, or what the original discount was. So here are a couple of ways to find out. Notice that each discrete quantity has been put in parentheses, just to set it apart as representing one quantity.

If you look at it from a logical point of view, to find the discounted amount, multiply the percent discount by the original price and then subtract that amount from the original price:

$$\text{(Discounted price)} = \text{(Original price)} - [\text{(Original price)} \times \text{(Percent of discount)}]$$

Another way to look at this—and perhaps an easier one to deal with— says to subtract the amount of the discount from 100% and multiply the original price by that difference:

$$\text{(Discounted price)} = [100\% - \text{(Percent of discount)}] \times \text{(Original price)}$$

Using the formula we just spelled out, for reasons that may or may not be apparent, if any two of the quantities in the above relationship are known, it is possible to find the third. That means that if we know the discounted price and the percent of the discount, we can solve for the original price.

$$\text{(Original price)} = \frac{\text{(Discounted price)}}{[100\% - \text{(Percent of discount)}]}$$

Also, if we know the original price and the discounted price, we can find the percent of the original price the item is being sold for and the percent of the discount.

$$\text{(Percent of original price)} = \left[\frac{\text{(Sale price)}}{\text{(Original price)}}\right] \times 100$$

To find the original percent discount, subtract the (Percent of original price) from 100%:

$$\text{(Percent of discount)} = 100\% - \text{(Percent of original price)}$$

Whew! That's a lot to remember, but you really don't have to. There's a "Figuring Discounts" page in the back of this book with all the formulas, and you can just plug the numbers into the formulas to find the answer, just like you can do with the problems below.

Pop Quiz

1. An item was discounted 25%. Another 25% was taken off the discounted price. What percent of the original price is the item now selling for?

2. An item is selling for $35 after it was discounted by 30%. What was its original selling price?

3. An item's list price is $250. It is selling for $200. How much was it discounted?

Answers

1. If the item sold for $100, then a discount of 25% would bring the price down to $75. Next, we take 25% off of the $75 by taking 10% of it twice and then 5% of it.

   That's $7.50 + $7.50 + $3.75 = $18.75.

   $75 – $18.75 = $56.25. Remember the magical connection between percent and $100. The item is selling for 56.25% or $56\frac{1}{4}$% of its original price.

2. To solve this one, plug the known values into the formula above for finding the original price.

   $$(\text{Original price}) = \frac{(\text{Discounted price})}{[100\% - (\text{Percent of discount})]}$$

   We express percent as a decimal, since that's the easiest way to work with it, although you could have used 100/100 and 70/100 in place of 1.00 and 0.70 had you preferred.

   Substitute: $(\text{Original price}) = \frac{35}{(1.00 - .30)}$

   Subtract: $(\text{Original price}) = \frac{35}{.70}$

   Divide: $35 \div .70 = 50$

   The original price was $50.

3. We start solving this one by using the next-to-last formula stated previously.

$$(\text{Percent of original price}) = \left[\frac{(\text{Sale price})}{(\text{Original price})}\right] \times 100$$

Substitute: $(\text{Percent of original price}) = \left[\frac{(200)}{(250)}\right] \times \frac{100}{1}$

Multiply: $(\text{Percent of original price}) = \left[\frac{(20000)}{(250)}\right]$

Divide: $(\text{Percent of original price}) = 80$

That's 80%, meaning the sale price is 80% of the original price. The discount then was 100% – 80%, or 20%.

7

Measurement

There are essentially two systems of measurement in general use in the world. One of those is very logical and organized; it is used by most of the world. The second is very illogical and filled with potential for confusion. It is the one used in this country. The logical and organized system was invented in France during the Napoleonic era. It was originally known as the **metric system,** which has been refined to the modern **Systeme Internationale** (said with a French accent, and often abbreviated **SI**—pronounced Ess-Eye).

The system of measurement we use in the United States is known as the **English system,** or **traditional measurement.** This is somewhat ironic, since the English no longer use it. It was, at one time, standardized based on some of the body parts of King Henry VIII.

Linear Measure

Linear measure is the name we give to measuring distance—that is, measuring in a straight line or in one dimension. Whether you're building something out of wood or measuring to put a fence around your yard, you'll need some device to measure distance. The three most common distance-measuring devices found around the home are the ruler, the yardstick, and the measuring tape. All three devices are marked in **inches** and fractions of inches (halves, fourths, eighths, and sixteenths, usually). If the device you're using is longer than twelve inches, it is probably also marked in **feet.** One foot (abbreviated ft.) equals twelve inches (abbreviated in.). Sometimes the periods are omitted from those abbreviations, so that you might write:

$$1 \text{ ft} = 12 \text{ in}$$

If you happen to have a plastic (or metal) one-foot ruler handy, you've probably already noticed the little numbers on the long edge opposite the inch markings. Those are **centimeter** markings. About 30.5 cm (short for centimeters and never written with a period) are equal to one foot. We'll talk more about metric units in a bit.

Part of a foot-long ruler

While you'll find yardsticks around most paint and hardware centers, you probably remember the yardstick best from school. Whether you remember it fondly or not will probably depend on whether you used it to create figures on the chalkboard or the teacher used it to make marks on you. A yardstick contains three feet, or is exactly thirty-six inches in length. Two yards of length (six feet or seventy-two inches) is equal to one **fathom,** the rather well-known nautical measurement.

Do you see the logic of this system of measurement? Oh, yes; there're two or three more to consider:

$$5280 \text{ feet} = 1 \textbf{ mile}$$

That's the same as 1728 yd. (yards) or 864 fa (fathoms). Don't even think about remembering the last two. You'll never need them. I just threw them in to show you what an illogical system we're dealing with. Twelve of these equal one of those but three of those equal one of the next thing, then twice that for the next and 864 of those for the last. If the last sentence didn't make sense to you, consider substituting the following words for "these," "those," "the next thing," "the next," and "the last": "inches," "foot/feet," "yard(s)," "fathom(s)," "mile," in that order.

Now take a deep breath and get ready for some fresh air.

SI Units

We have spent nearly two full pages talking only about linear measure in the traditional system. We haven't touched weights or liquid capacity. I'm now going to knock off all of that in SI units in about a single page. Remember, SI is roughly identical to the metric system.

In SI units, the unit of linear measure is the **meter,** roughly equal in length to the yard (actually, just a little longer). The unit of **mass** (which we'll treat as interchangeable with weight, though it's technically not),* is the quite small **gram.** The unit of liquid capacity is the **liter,** slightly larger than a quart, which you may be accustomed to thinking of as half a two-liter soda bottle.

Within any type of measure, the SI unit is prefaced by a prefix based on multiples of ten, as follows:

Prefixes for SI Units
milli- means × 1/1000
centi- means × 1/100
deci- means × 1/10
no prefix means the unit itself
Deka- means × 10
hecto- means × 100
kilo- means × 1000

You may be wondering why Deka- is capitalized. It's to distinguish "Dm" (Dekameter) from "dm" (decimeter).

That's it. A **kilogram** is the most often-used unit of weight, being equivalent to about 2.2 pounds. A **kilometer** is about 0.6 of a mile, and there are about 240 **milliliters** in a typical 8 fluid ounce cup of water. There are larger and smaller SI prefixes to take care of whatever needs the user might have. And of course if you don't use them every day, SI units take some getting used to, but I think you'll agree that the system is elegant.

Cutting the Rug

To "cut a rug" is an old-time expression meaning to "dance," but it isn't being used that way here. If you're going to buy carpeting, or even an area rug, you need to be able to measure the area that you're planning to cover, and you don't buy carpeting in linear feet or yards. Carpet and other floor coverings are sold in square units, known as **units of area** (or two-dimensional units).

---

*Mass is the amount of matter in something. It is the same wherever that thing is. Weight is a force (mass times the acceleration of gravity). Where a = the acceleration of gravity, m is mass, and f is weight, $f = ma$. Now there's a footnote! It's also known as Newton's Second Law of Motion.

Determining Area

Let's start out with the most basic of floor plans—the rectangular room.

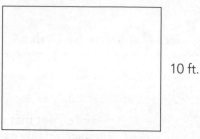

13 ft.
10 ft. by 13 ft. room

Suppose you wished to cover the floor of this room with 12 in. by 12 in. carpet tiles. How many of them would you need? A 12 in. by 12 in. carpet square is more commonly known as a **square foot tile.** Its measure is written as 1 ft.^{2} When something is written with an **exponent** of 2 after it, that is an indication that it is being multiplied by itself. One ft.^{2} means one foot has been multiplied by one foot. To see why, go back to the question that began this paragraph. Then look at the figure. To find the area of the rectangular room we multiply its length times its width.

$$10 \text{ ft.} \times 13 \text{ ft.} = 130 \text{ ft.} \times \text{ft.} = 130 \text{ ft.}^2$$

Just as the two numbers are multiplied together, the two dimensions are also multiplied together. Look at the next figure. The lines have been drawn in so you can count the number of square units. Go ahead, count them.

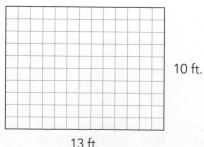

13 ft.
The same room, tiled

Did you count the 130 tiles? Which way was easier, counting them or multiplying length times width? Don't worry, I knew the answer. It took a lot less time that way, too. The surface of any flat figure is expressed in units of area. Similarly, carpet is sold by area, so many dollars per square yard.

Before going on, let's develop a standard formula for the area of any rectangle. If we call the length of either horizontal component of the rectangle its base, represented by the letter, b, and the length of either vertical side its height, represented by the letter, h, then a rectangle's area, A, is found by the formula

$$A_{\text{rectangle}} = b \times h, \text{ or, more simply, } A_{\text{rectangle}} = bh$$

Suppose you wished to buy a style of carpet that was priced at $20 per square yard. Also suppose that you could buy exactly the amount that you needed. What would it cost you to carpet the room we just tiled?

The first question you need to answer is how many square yards there are in 130 ft.^{2} Before figuring that out, you need to know how many ft.^{2} = 1 yd.^{2} Remember there are 3 ft. in a yd. Look back at the preceding figure and mark off one square yard. It's 3 rows of 3 feet each, or 9 ft.^{2} So, if there are 9 ft.^{2} per yd.^{2}, we need to divide 130 ft.^{2} by 9. The quotient will not be a whole number. It's $14^4/_9$. Now find $14 \times \$20$ is 280; $^4/_9 \times \$20 = \$8.89$, to the nearest penny, so the cost would be $288.89.

Of course, you couldn't actually buy exactly the amount of carpet you needed, unless you have a room that is the exact same length or width as the width of the roll of carpet. As a general rule, you'll have to buy more carpet than you really need. You'd also have to pay for padding to lay under the carpet, and you'll probably want to pay for professional installation. Then there's sales tax, . . . Well, you get the idea.

The Fabric of Society

Not that many people work with fabric to sew their own garments these days, although drapes, curtains, and other accessories are frequently made by homemakers. Fabrics, like carpets, are also sold by the square yard, in fact, they are known as **yard-goods.** While you are unlikely to ever need to carpet a triangular-shaped room, there are many applications for triangular-shaped fabrics. Depending on how much joining together can practically be done, it makes sense to know how to find areas of some shapes other than rectangles.

The formula for the area of a right triangle is practically intuitive.

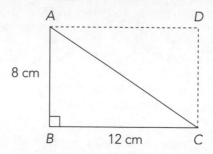

As you can plainly see in the figure above, the right triangle, $\triangle ABC$ (the $\triangle$ symbol identifies what I'm referring to as a triangle, and the *ABC* names its three vertices—corners) is formed by diagonal *AC* of rectangle *ABCD*. That diagonal cuts the rectangle into two equal-size triangles, $\triangle ABC$ and $\triangle ADC$, each of which has an area equal to half that of the rectangle. Since we already saw that the area of a rectangle is computed by the formula $A_{rectangle} = bh$, it makes sense to conclude that

$$A_{triangle} = \frac{1}{2}bh$$

For $\triangle ABC$, the area would be $\frac{1}{2} \times 12 \times 8$, or 48 cm^2. That's right! It may be the area of a triangle, but the units of area are still square. If you're not sure why, remember: We're still multiplying cm $\times$ cm. That makes cm^2.

The preceding argument makes sense for a right triangle, but what about an acute triangle (that is one where all angles are less than 90°)?

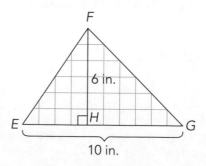

$\triangle EFG$ is an acute triangle with base *EG* and height *FH*. Note the height is not a side of the triangle. The height of any geometric figure must be perpendicular to (form a right angle with) the base. It certainly doesn't look to me like this triangle has an area that's half of some rectangle, but wait. Let's try something.

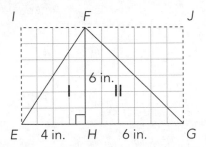

Here's the same triangle separated for convenience into $\triangle$I and $\triangle$II. I've completed a rectangle around each of those triangles. $\triangle$I is in rectangle *EHFI*; $\triangle$II is in rectangle *HGJF*. Both rectangles have the same height, 6 inches. The base of $\triangle$I is 4 inches long; that of $\triangle$II is 6 inches long. Notice that while triangles I and II make up acute $\triangle$*EFG*, each of them is a right triangle. That means that the $A = \frac{1}{2}bh$ formula definitely applies to each of them.

For $\triangle$I:	*For* $\triangle$II:
$A = \frac{1}{2}bh$	$A = \frac{1}{2}bh$
$A = \frac{1}{2}(4)(6)$	$A = \frac{1}{2}(6)(6)$
$A = \frac{1}{2}(24)$	$A = \frac{1}{2}(36)$
$A = 12$ in.^{2}	$A = 18$ in.^{2}

That means that the area of $\triangle$*EFG* = 12 in.^{2} + 18 in.^{2} = 30 in.^{2} Does that make sense to you?

Now go back and look at $\triangle$*EFG* without the rectangle. Suppose we applied the right triangle area formula to the acute triangle *EFG*, with base 10 in. and height 6 in.

$A = \frac{1}{2}bh$
$A = \frac{1}{2}(10)(6)$
$A = \frac{1}{2}(60)$
$A = 30$ in.^{2}

I'll be darned. Surprised? We didn't have to go through all the mumbo jumbo with the two right triangles and the two rectangles after all. The same formula that worked to find the area of the right triangle also works with the acute triangle. And now I'm going to give you one more piece of news. The same formula also works for finding the area of an obtuse triangle.

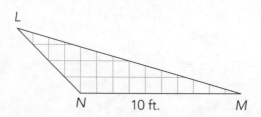

An **obtuse triangle** is a triangle that contains one angle whose measure is greater than 90°. Such a triangle is shown above. The trick in finding the area of an obtuse triangle comes in finding its height, which usually is going to fall outside the triangle.

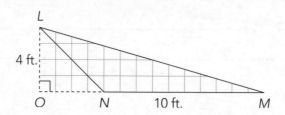

We extend side *MN* through *N* to *O*, and then drop segment *LO*, perpendicular to segment *MO*. *LO* is the height of △*LMN*. Now, to find the triangle's area,

$A = \frac{1}{2}bh$

$A = \frac{1}{2}(10)(4)$

$A = \frac{1}{2}(40)$

$A = 20 \text{ ft.}^2$

Pop Quiz
Find the areas of the pieces of cloth in 1–3:

1.

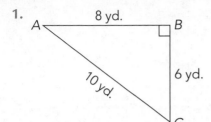

2.

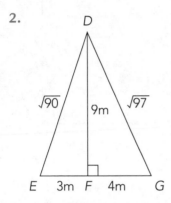

3.

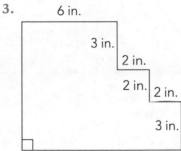

4. Examine the following floor plan. Then find how much it would cost to cover it with hardwood flooring at $6.95 per square yard.

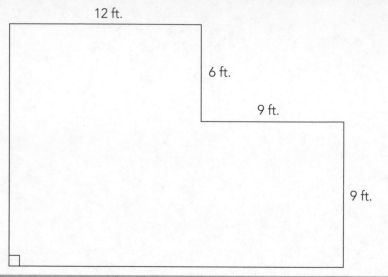

12 ft.

6 ft.

9 ft.

9 ft.

Answers

1. 24 yd.^{2}
2. 31.5 m^2
3. 64 in^2
4. $201.55

Round and Round We Go

While it's unlikely (but admittedly possible) that you'll ever have to tile a round room, you well may wish to buy a round area rug, and you are very likely to wish to cover (at some time or another) a round table, or to place chairs around a round table. The basic round plane figure is the circle, and there are many line segments associated with a circle, only two of which you need to concern yourself with.

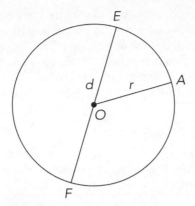

PARTS OF A CIRCLE

A circle is named by its center, so the above figure shows Circle O. Segment *OA* connects the center of the circle to the **circumference** (the name given to the outside—or perimeter) of the circle. A segment that connects the center to the circumference is known as a **radius.** A circle, in theory, may have an infinite number of radii (plural of radius). The longest distance across a circle is the circle's **diameter.** Segment *EF* is a diameter of circle O. Notice that *EF* consists of two radii.

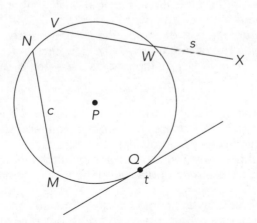

Circle *P* is drawn as a service to the curious. It shows the other segments associated with a circle that you have no need to know. Segment *MN* connects two points on the circumference. It is known as a **chord.** A circle's diameter is also its longest possible chord. *VX*, a chord that continues beyond the circle, is known as a **secant.** *VX* happens to have an endpoint at outside point *X*, but a secant may be infinite in length. The same is true of the **tangent** to *Q*. It touches a circle at one point only. I could tell you more about that line's properties, but I'll refrain from going off on that tangent!

A PIZZA PIE (PIECE OF PI)

The relationship between the length of the diameter of a circle and the length of its circumference was discovered thousands of years ago by the ancient Greeks.

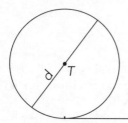

They found that if a wheel were rolled along the ground for one full turn, the distance that it traveled would be 3 and a little bit more than the length of its diameter. They gave that relationship a name, after the Greek letter, π (pi). But π is an irrational number that can only be approximated. It is customary to use $^{22}/_7$ or 3.14 as the approximation of pi, for purposes of computing circumferences and areas of circular regions. The circumference, *C*, of any circular figure or circular object can be found by multiplying its diameter, *d*, by π:

$$C = \pi d \qquad \text{or} \qquad C = 2\pi r$$

The 2 and the *r* replace the *d* in the second formula, since the diameter is equal in length to twice the length of the radius. Need a skirt for a round bed? Now you know how to figure out the length of the material you should buy. You can also compute the amount of edging you'll need to decorate around a round table or to trim a festive pie plate.

But wait; we still haven't come full circle. You still need to be able to find the area of a round region to figure out how much cheese spread you'll need for those round canapés you were planning to serve your company. For the area of a circle, use this formula:

$$A = \pi r^2$$

That's 3.14 × the radius × the radius, and yes the answer will be in square units. Consider a sixteen-inch-diameter pizza. How many square inches of mozzarella will be needed to cover the entire pie with cheese?

$A = \pi r^2$

$A = (3.14)(16)^2$

$A = (3.14)(16)(16)$

$A = (3.14)(256)$

$A = 803.84 \approx 804$ in.^{2}

Pretty cheesy, eh?

Pop Quiz

With the understanding that π is 3 and a little more, you may leave your answers in terms of π, where appropriate.

1. A chalk mark is at the bottom of a 17-inch-diameter tire. If that tire turns exactly nine times in the same direction, how far will that chalk mark travel?

2. A round tabletop is 81π ft.^{2} in surface area. What is the distance across it at its widest part (or diameter)?

3. A gallon of paint covers 150 ft.^{2} How many gallon cans would you need to cover a round deck with a diameter of 14 yd.?

Answers

1. 153π in.

2. 18 ft.

3. 10 cans

Did I trick you with Question 2? If the tabletop's area is 81π ft.^{2} and area is πr^2, then the radius, r is 9 ft. But the distance across the table is its diameter, twice the radius, or 18 ft.

How about Question 3? A deck with a diameter of 14 yards has a radius of 7 yards. Since we were given coverage in square feet, we must change 7 yards to 21 feet (3 ft. per yard). Then we find 21 squared is 441. Multiply using 3.14 for π. That makes 1384.74 ft.^{2} Now divide that by 150, since that's what one gallon covers and you'll get 9 and a little bit more. Well, you can't leave part of the deck unpainted, and you can't buy part of a can of paint, so you'll need a tenth can for the extra paint, and the answer is 10.

Trounced by the Ounce

If the heading of this section makes no sense to you, consider the following. The ounce is the smallest normally used weight in the traditional system of measures. It is $^1/_{16}$ pound. In other words, 16 **avoirdupois** ounces make one pound. "Avoirdupois?" Yep, that's what they're called. It's from the Old French, *aveir de peis* meaning literally "goods of weight," referring to things sold by weight (rather than by the piece). The ounce, pound, and ton are really the only traditional weight measures in common use in this country. There is, however, also the **troy** ounce, which is used by apothecaries and is equal to $^1/_{12}$ pound.

But weight—er, wait—there's more! Our measure of fluid capacity is usually the fluid ounce, which is $^1/_{32}$ quart. You're not likely to come up against troy ounces every day, but you'll deal with things measured in fluid ounces and avoirdupois ounces practically every day of your life. The British also deal with stones, which are currently defined as a weight of 14 pounds, so if a person weighs 10 stone, (s)he weighs 140 pounds (abbreviated lbs.).

The ounce (weight) is abbreviated oz., while the fluid ounce is abbreviated fl. oz., with the periods being optional. You'll also come across things that you'd expect to be sold by liquid capacity being sold by weight. For example, sweet cream is sold by the pint (16 fl. oz.), but sour cream is often sold by the pound. That should be enough to knock you on your dairy-erre!

Fluids

Let's take a look at fluid measures first. Fluid means something that flows, and includes gases as well as liquids. Since the only gases we normally deal with are in the atmosphere or in the line or tank from the gas company, we're really not going to pay that type of fluid much attention here.

The main commercial measure of liquids is the fluid ounce and its specially named multiples. By specially named multiples I mean the cup, pint, quart, and gallon. The following table tells the story.

Measure	How Many?	Fluid oz.
cup (c)		8
pint (pt.)	2 cups	16
quart (qt.)	2 pints	32
gallon (gal.)	4 quarts	128

You may well ask why you need to know these measures. The most important answer is for smart grocery shopping. Many things are sold in pints, quarts, half-gallons (which is not an actual unit of measure, but you should be able to figure out what it equals), and gallons— especially in the dairy case.

There are other units of liquid measure, like the drop, the minim, the fluid dram, and the gill, but none of those are in common use today, and there is no reason why you need to know them, so, save for this passing mention of their existence, we shall avoid dealing with them. I shall include them, however, on the Measurement Tip Sheet in the appendices.

As a general rule, but not a true and fast one, the larger the container you buy, the less you pay per fluid ounce. While the last statement is not always true, it is true enough of the time, so that you should realize that it usually pays to buy the largest sized package of a product that you and your family can consume in a reasonable amount of time. When some smaller than the largest size packages of a product are on sale the bigger is cheaper rule may not apply. Then, it may pay to buy two or more of the smaller packages at the sale price.

Weighting for Godot

The main units of weight in the traditional system are the ounce, the pound, and the ton. Sixteen ounces (abbreviated oz.) equal one lb. (pound). Two thousand pounds equal one ton (T.). Since I doubt that you'll ever buy anything that's sold by the ton (unless you're in the construction business), the first two measures are the only ones you need to know. If you happen to live in the U.K., you also ought to know that fourteen pounds = one stone. We'll look next at the practical aspects of both this and the last sections.

Comparison Shopping

It has become easier to comparison shop these days than in those of your parents, at least at the supermarket, where laws require the posting of unit prices. Certainly, you've noticed labels like the one in the following figure.

```
┌─────────────┬──────────────────────────┐
│ UNIT PR.    │  YUMMIES CORN FLAKES     │
│   .11/oz    │                          │
├─────────────┴──────────────────────────┤
│                                         │
│        12 oz. Box $1.32                 │
│                                         │
└─────────────────────────────────────────┘
```

The store trumpets the name of the product and the cost of the box. The unit price may be in the upper left-hand corner of the shelf tag, the lower right-hand corner, or somewhere on that tag, and usually in tiny print. To the store, the unit price is not the most important thing. To you, the customer, it should be. Before you buy those brand name Yummies Corn Flakes (at 11 cents per ounce), find out what the store brand's corn flakes' unit price is. There's often little or no difference in quality. Indeed, sometimes they're both made by the same manufacturer, but the price of the name brand product includes charges to cover what they paid to advertise it. To get the best buy, try the store's brand. If you don't like how it tastes, don't buy it again. Keep track in a notepad of what off brands you've tried and whether or not you liked them. Sometimes you can save as much as one-third by buying store brands. That can mount up to big bucks in a short time.

Unit pricing makes it easy to compare the real cost of what you are buying. The supermarket doesn't play up the unit price, but if you saw a 16-ounce can of tomato soup for $1.28 with a unit price of .08/oz. and a 19-ounce can of tomato soup for $1.52 and a unit price of .08/oz., you are not saving by buying the larger can. Both soups cost exactly the same eight cents per ounce. Buy the larger can if you need more soup. If 19 ounces is more than you need, buy the smaller can and save $0.24 to spend on something else.

Beware of Downsizing

An unsavory but widespread trend among manufacturers in recent years is to keep the same-size package and the same product price but put a smaller quantity of product inside the package. This practice has been most widely used with paper products, cleaning products, baby food, dry cereals, ice cream, snack foods, and coffee.

You may be in the habit of grabbing your most commonly used products and making a dash for the checkout line. Stop! Take a few extra moments to check out how much product is actually inside the package. That way you can be sure you are getting what you think you have bought in the past.

Keeping a small notepad with information on your favorite products is a very handy way to keep track. Downsized packages are specifically designed to be unnoticeable to most shoppers. Don't be fooled by them.

Pop Quiz

1. A 12 fl. oz. bottle of soda pop costs $0.96. A pint of the same pop costs $1.28. Which is the better value?

2. A 3-pound bag of russet potatoes costs $1.83; A 5-pound bag of Yukon Gold potatoes costs $2.90. Which is the better value?

3. A quart of milk is $0.89. A gallon of milk costs $2.96. How much money is saved by buying one rather than the other?

Answers

1. Neither; they both cost $0.08 per fl. oz.

2. The russets cost $0.61/lb. The Yukon Golds cost $0.58 per pound. They are the better value.

3. Four quarts make a gallon, so divide $2.96 by 4 to find the price of each quart it contains: $2.96 ÷ 4 = $0.74. By buying the gallon you'd save $0.15 per quart; ($0.89 – $0.74 = $0.15).

8

Geometry, Plane and Not So

The reason for this chapter's existence may not be obvious at first, but bear with me. Geometry has many everyday uses. Plane geometry is the math for people who never liked math. Why? Because you don't need to use algebra. Plane geometry is the study of **plane figures**— that is, figures on a flat surface. Plane figures have length and width, which in certain cases are the same. They have no depth. Figures with depth are called **solids,** and we'll look at them toward the end of this chapter. To be perfectly honest, we already did some plane geometry in Chapter 7, when we dealt with areas, but we did it in sort of a vacuum, where we were dealing just with the concept of surface measure.

Basic Concepts

There are some terms that relate specifically to geometry. You probably already know them, but you may not know their strict definitions. For example, the basic unit of geometry is the point. I mean that literally, as in the following figure.

Point A

A **point** is a location on a plane or in space. For the moment, we'll be concerned only with the former. A point occupies no space. It is represented by a dot on a piece of paper, and is named by a single upper-case letter, hence Point A, as shown in the figure. The dot representing Point A is actually bigger than the point itself. An actual

100

point, I'll remind you, takes up no space, but if Point A were drawn taking up no space, you wouldn't be able to see it.

The next basic geometric entity is the **line,** represented in the figure that follows.

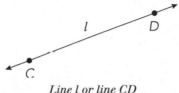

Line l or line CD

A line is an infinite series of points and extends forever in both directions, as indicated by the arrowheads. It may be named by a single lower-case letter, like line *l*, or by a pair of points on it, as $\overleftrightarrow{CD}$. A line has no direction, and so $\overleftrightarrow{DC}$ would be another name for it, although it is customary to say the letters in alphabetic order (a custom—not a rule). Any line is considered to be straight, so the expression "straight line" in geometry is redundant.

Another common phrase that is incorrect in geometry is "a line is the shortest distance between two points." Since a line is endless, the only two points it could connect would be positive and negative infinity, which are by their nature concepts rather than locations. The shortest distance between two points is actually a **line segment.**

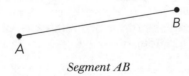

Segment AB

The figure above represents line segment *AB*, often represented as $\overline{AB}$. That distinguishes it from $\overleftrightarrow{CD}$, which represents the line in figure "Line *l* or line *CD*." A line segment, as you've figured out by now, is named by its two endpoints. Again, order doesn't matter.

And now for something completely different. A **ray** may be considered to be half a line, but don't quote me on that. It has similarities to both segments and lines. Like a segment, it has an endpoint; like a line, it goes on forever—but only in one direction. The figure below shows ray *LM*, a.k.a. $\overrightarrow{LM}$.

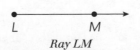

Ray LM

Unlike a line or line segment, a ray has direction: $\overrightarrow{LM}$ is not the same as $\overrightarrow{ML}$ since it goes the other way. A ray is named by its endpoint and some other point on it.

Getting a New Angle

Imagine $\overrightarrow{AB}$ lying directly on top of $\overrightarrow{AC}$ so as to appear to be a single ray. Next, we grasp $\overrightarrow{AB}$ and rotate it around A about 30°.

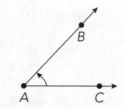

Acute angle A, or BAC, or CAB

What we get is an **angle,** which is defined as two rays with a common endpoint. That common endpoint is known as the angle's **vertex.** An angle may be named by its vertex, with the symbol ∠ standing for the word "angle." Here we have ∠A. It may also be named by letters from its surrounding sides and the name of the vertex, so this may be ∠BAC or ∠CAB. There is a third way to name angles. A numeral may be written inside the angles, thus designating them ∠1, ∠2, etc.

Depending on whom you ask, you'll be told that there are three different types of angles or four different types of angles. They are shown in the following figure.

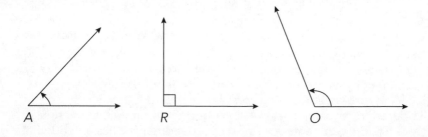

The three or four types of angles

All of these angles are drawn in what is known as **standard position.** That means that each is drawn with its initial side horizontal and to the right, and its vertex on the origin (where the x-axis and y-axis from Cartesian Coordinates* cross), and with the terminal side (the side that has been rotated) elsewhere.

In the above figure, ∠A is **acute,** the name given to all angles less than 90°, while ∠R is a **right** angle, equivalent to one-quarter of a complete circular rotation, or exactly 90°. The corners of this book form right angles, so keep it handy for testing purposes.

Angle O is **obtuse.** That doesn't mean it's difficult to understand. An obtuse angle is one in which the terminal side has been rotated more than 90° and less than 180° (or a half circle's worth).

Finally there's the fourth type of angle, which some refuse to acknowledge as an angle. It's ∠S, formed by two opposite rays. This **straight angle,** ∠S, could really be considered as two right angles placed back to back with the vertical ray(s) removed. That's why the measure of a straight angle is 180°. Consider what would happen if the terminal side of a straight angle were swept through another 90° to 270° and then yet another to come back to where it had started. It would have rotated through a complete circle, or 360°. And that is why a circle is considered to have a degree measure of 360°; but I digress.

When the sum of the measures of two or more angles totals 90°, those angles are said to be **complementary**—spelled with two e's. When the sum of the measures of two or more angles totals 180°, those angles are said to be **supplementary.** Now you know all the angles.

The Parallel Universe

There are essentially two types of lines, ones that intersect and ones that don't.

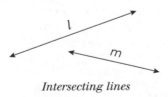

Intersecting lines

---

*Cartesian Coordinates: A rectangular grid with x-values in the horizontal direction and y-values in the vertical direction, named after French mathematician and philosopher Rene Descartes. Each point is assigned coordinates (x, y).

Does the name of the above figure surprise you? They converge, you say, but they don't intersect. I told you geometric terms don't always agree with everyday ones. Lines *l* and *m* may not appear to intersect, but remember that lines are infinite in length. Even though the two representations of the lines don't cross, since both can be extended indefinitely, they will cross. Therefore they are intersecting lines. **Intersecting lines** cross at one point and one point only. No matter how much beyond that intersection those lines are extended, they will never meet again.

Some lines never meet.

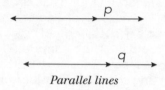

Parallel lines

Lines *p* and *q* are parallel lines. You can symbolize this by writing *p//q*. **Parallel lines** never intersect, regardless of how far they're extended. The arrowheads near the middle of the diagrammed lines indicate parallel lines.

Angles formed at intersection

When lines or line segments intersect as do $\overline{AB}$ and $\overline{CD}$, angles are formed at the intersection. Angles formed on opposite sides of the vertex (point of intersection) are referred to as **vertical angles.** That means on opposite sides of the vertex, not up and down. In the above figure, ∠1 and ∠3 are a pair of vertical angles. So are ∠2 and ∠4. Vertical angles are equal in degree measure. That's symbolized as follows:

$$m\angle 1 = m\angle 3 \text{ and } m\angle 2 = m\angle 4$$

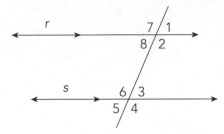

A transversal cutting two parallel lines

When two parallel lines are cut by a **transversal** (a line that intersects both of them), as in the figure above, certain angles are formed. I'm going to try to get all of these in on one diagram so I don't have to use up more space than I've been allotted. Angles 1 and 3 are called **corresponding angles of parallel lines.** They are located in the same position with respect to each line and the transversal. They are always equal. Conversely, if two lines are cut by a transversal so that their corresponding angles are equal, that is adequate proof that the lines are parallel. See how many other pairs of corresponding angles you can find in the above figure. I'll wait. Jot them down. Then you can look on the bottom of the page.*

Notice that there are two sets of vertical angles at the intersection of line *r* with the transversal and two more sets at the intersection of line *s* with the transversal. There are also four sets of supplementary angles at each of those intersections. There are some other relationships of equality and supplement that exist between pairs of angles at both intersections, but studying them is really more of an academic activity than a useful one.

That Figures

Closed figures are figures with sides consisting of line segments (not counting circles, of course), which have an inside and an outside separated by a **perimeter** (distance around). Such figures (again, except for circles) are known as **polygons,** meaning many sides. The fewest number of sides needed to make a closed figure is three.

As I'm sure you've figured out by now, the simplest (smallest number of sides) closed figures are more popularly known as **triangles.** Triangles may be classified according to their angles and/or according to their sides. Let's take the angles first.

---

*∠2 and ∠4; ∠6 and ∠7; ∠5 and ∠8.

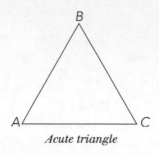

Acute triangle

The above figure depicts △*ABC*, an **acute triangle.** It is so called because every one of its three angles is acute, which we've already defined. You can remember it that way, or just look at it and say to yourself, "My, what a cute triangle!"

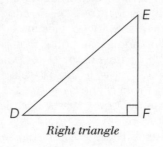

Right triangle

The figure above shows △*DEF*, which is one possible form of a **right triangle.** It is named by the one right angle that it contains. It doesn't matter how the triangle is oriented. If it contains a 90° angle, there's just something right about it. Note that two of the angles in a right triangle are acute.

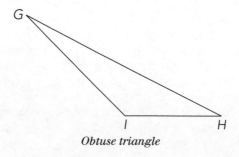

Obtuse triangle

You've probably figured out the third type of triangle by angle classification, and it is depicted in the previous figure as △*GHI*. This is an example of an **obtuse triangle.** Again, both angles other than the one obtuse angle are acute.

Next we're going to classify triangles by their sides, but keep in mind that the side classifications affect relationships among the triangles' angles.

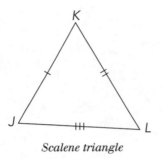

Scalene triangle

The above figure shows △*JKL* as a triangle with sides of three different lengths. Such a triangle is known as **scalene.** Because the sides are all of different measure, so are the angles. In fact, the angles of any triangle are related to the sides they are opposite in the same proportion. To tell which side of a triangle is opposite which angle, pick an angle and consider it to be a pair of gaping jaws. The side in the teeth of those jaws is the opposite side, so side $\overline{KL}$ is opposite ∠*J*.

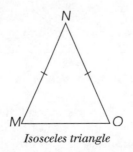

Isosceles triangle

The above figure shows △*MNO* as an **isosceles triangle.** That means two of its sides are equal in length. Notice that sides $\overline{MN}$ and $\overline{NO}$ are both marked the same way. That means that they are equal in length. But what does that mean for the angles opposite those two sides, the base angles ∠M and ∠O? Well, if you haven't already guessed, it means that they are equal, too.

By the way, the vertex where the two equal sides meet is known as **the vertex of the triangle.** I don't think you really need that information, but it's true. Every triangle has three vertices (vertexes), but only an isosceles triangle has the vertex of the triangle.

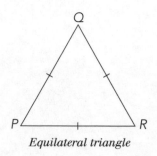

Equilateral triangle

Could you tell what was coming? Yes, △PQR is a triangle with all three sides equal in length. It is known as an **equilateral triangle.** Without any further ado, an equilateral triangle is also an **equiangular** one.

Suppose you were to draw a triangle on a piece of paper, and then cut it into three parts, each of which contains one of the angles. Then you laid those pieces of triangle next to each other so that all three vertices were at the same point, as in the figure below.

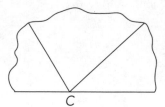

The sum of a triangle's angles

Notice that point C is the vertex for all three angles. Notice that the three angles together form a straight angle. That is, any one of those angles is **supplementary** to the sum of the other two. Then it would be correct to assume that the measures of the angles of any triangle sum to 180°. If you don't believe me, go ahead and try it. Heck, try it whether you believe me or not.

Pop Quiz

Answer each question in the most appropriate way.

1. What is the name given to a triangle with no sides equal in measure?

2. What is the name given to a triangle with one angle greater than 90°?

3. What is the sum of the degree measures of two complementary angles?

4. The sum of the measures of all angles of a triangle must add up to ___°.

5. What is the name given to a triangle with one 90° angle and two angles with exactly 45°?

Answers

1. scalene
2. obtuse
3. 90°
4. 180
5. right isosceles (or isosceles right)

Ratio and Proportion

A **ratio** is a comparison. The sides of a 4-inch by 6-inch photograph are in the ratio 4:6, read "4 is to 6." A ratio may also be written as a fraction (as mentioned in Chapter 4), so $\frac{3}{5}$ may be read as "3 is to 5." Ratios are really not of much value by themselves, but when equated (set equal to one another), they become extremely useful. An equating of two ratios, such as 3:6 = 1:2, is known as a **proportion.** Needless to say, that can also be written as $\frac{3}{6} = \frac{1}{2}$. When a proportion is written in the colon format, the two numbers closest to either side of the "=" sign are known as the **means.** The numbers farther away from the equal sign are the **extremes.** The means and extremes are not quite as obvious when a proportion is written in fractional form, but the means are the denominator of the first fraction and the numerator of the second. That makes the extremes the numerator of the first fraction and the denominator of the second. If a proportion is **true,** the product of the means equals the product of the extremes. Go ahead and test the proportions above to see whether they're true.

Using the property of a proportion that we just described, it becomes a simple matter to find the fourth member of a proportion when the other three members are known. Consider the following:

$$8:12 = 2:\_\_$$

What goes in the blank? To find the answer, multiply the means together and get $2 \times 12 = 24$. If the product of the means is 24, then the product of the extremes must be 24. What times 8 equals 24? You might also read that as 24 divided by 8 equals what? The answer, of course, is 3.

In two triangles, if two angles of the first have the same degree measure as two angles of the second, then the third angles of each must also have the same degree measure. Two such triangles are known as **similar triangles.** They are identical in shape, but not necessarily the same size.* There's no need for the triangles to have the same orientation to be similar; only for them to have angles of equal measure.

The following figure shows two similar triangles, $\triangle ABC$ and $\triangle DEF$.

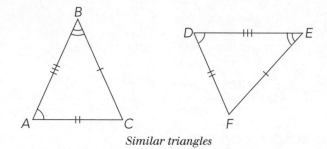

Similar triangles

Notice that $\angle A$ corresponds to $\angle D$, $\angle B$ corresponds to $\angle E$, and $\angle C$ corresponds to $\angle F$. The sides opposite those corresponding angles are known as the corresponding sides of two similar triangles. Side $\overline{AB}$ corresponds to side $\overline{DE}$, and so forth. Now here's the really cool part. Corresponding sides of similar triangles are in proportion. That means $\overline{AB} : \overline{DE} = \overline{BC} : \overline{EF}$, $\overline{BC} : \overline{EF} = \overline{AC} : \overline{DF}$, and $\overline{AB} : \overline{DE} = \overline{AC} : \overline{DF}$. In addition to the corresponding sides being in proportion, any pair of sides in one similar triangle are proportional to the corresponding pair of sides in the other. That is, $\overline{AB} : \overline{AC} = \overline{DE} : \overline{DF}$, etc. Any and all of those proportions are also reversible. "So what does that have to do with the price of coffee?" you might well ask.

---

*Two triangles that are identical in shape and size are called "congruent triangles," but they have no practical value.

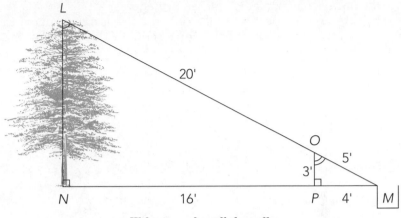

Wake up and smell the coffee

In the above figure there is a coffee tree, along which I have drawn line segment $\overline{LN}$, perpendicular to the ground. I'd like to know the height of this coffee tree, so I have measured 16 feet from the center of its trunk (don't ask me), and I have placed a yardstick ($\overline{OP}$) that stands up on its own fold-out base, attached for just such an occasion. Next, I get into the trench that was conveniently there and sight both the top of the yardstick and the top of the tree from ground level at point M, effectively creating two similar right triangles. How do you know that the triangles are similar, and what two triangles am I talking about? The two triangles are $\triangle LMN$ and $\triangle OMP$. Since $\angle O$ is in both triangles and both triangles contain a right angle, by definition they are similar. All that is needed now is to form a proportion between two of the known sides and the one we're looking to find. $\frac{MP}{MN} = \frac{OP}{LN}$. Filling in the known numbers, we get $\frac{4}{20} = \frac{3}{LN}$. Simplify the first fraction to get $\frac{1}{5} = \frac{3}{LN}$. Multiply the means and the extremes to get $1 \times \overline{LN} = 3 \times 5$. That gets you $\overline{LN}$. The coffee tree is 15 feet tall. Ta da! And you can even do this at home!

The Pythagorean Theorem

You may be wondering what all this triangle stuff has to do with everyday math. Well, it's background material for some practical uses that are coming up. It may be obvious to you that a right triangle is half of a rectangle cut by its diagonal, but it might not be obvious to anyone but the ancient Greek mathematician Pythagoras that the square on the **hypotenuse** (longest side) of a right triangle is equal in area to the sum of the squares of the two **legs** (shorter sides that form the right angle).

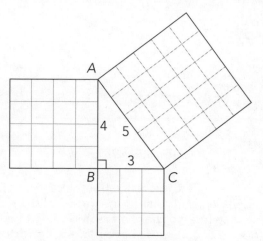

The Pythagorean Theorem

The figure above shows this relationship, which has for thousands of years been known as Pythagoras's Theorem, or more often as the **Pythagorean Theorem** (a proposition that is not self-evident, but can be proved).

The triangle in the figure is known as a 3-4-5 right triangle—a rather special triangle we'll discuss more fully in a moment. Note that the area of the square on the 3 side is 9 square units, and the area of the square on the 4 side is 16 square units. Add those together, and you get 25, the area of the square on the hypotenuse. This is true of all right triangles.

This relationship can also be written as

$$a^2 + b^2 = c^2$$

where *a* and *b* are the lengths of the two legs of the triangle, and *c* is the length of the hypotenuse. What that means is that given the lengths of any two sides of any right triangle, you can always compute the length of the third.

Unfortunately, with most right triangles, if you put in two sides and compute, you are likely to get a weird number for the third. For instance, suppose we had a triangle whose legs were 4 inches and 5 inches. Putting that into the formula, we get

$$a^2 + b^2 = c^2$$
$$4^2 + 5^2 = c^2$$
$$16 + 25 = c^2$$
$$c^2 = 41$$

Taking the square root of each side we find: $c = \sqrt{41}$.

Well, what the heck is the **square root** of 41? In other words, what number multiplied by itself makes 41? Ya got me! I know the square root of 36 is 6, and the square root of 49 is 7, so the square root of 41 must be between 6 and 7, probably closer to the 6 than the 7. Let's call it about 6.4 (my scientific calculator says 6.403124237). Not a very satisfying answer.

Suppose you know that the hypotenuse of a right triangle is 15 cm long, and one leg is 12 cm. Do you know how to find the other leg? Use the formula, but this time you know the hypotenuse, so

$$a^2 + b^2 = c^2$$
$$12^2 + b^2 = 15^2$$
$$b^2 = 15^2 - 12^2$$
$$b^2 = 225 - 144$$
$$b^2 = 81$$

And what number times itself makes 81?

$$b = 9$$

Wow! That worked out perfectly. Do you think it was an accident? Fahgeddaboudit! Remember the 3-4-5 triangle from back in the "Pythagorean Theorem" figure? Well, multiply each of those numbers by 3 and you get a 9-12-15 triangle; the same as a 3-4-5 but with sides bigger by a multiple of 3.

Triangles that work out perfectly (so that the sums of the squares of two sides make a square that has a square root that's a whole number) are known as **Pythagorean Triples.** Other common examples of Pythagorean Triples are 5-12-13 triangles, 8-15-17 triangles, and 7-24-25 triangles. You're more than welcome to prove to yourself that they work.

I opened this section by telling you that it had a practical use. Now I'm going to tell you what that is. Whether you're making a garden, laying out the foundation for a house, or stringing a fence, you want to get your corners perfectly square—that is, at right angles.

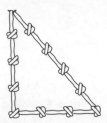

A knotty solution to a knotty problem

Some builders use a knotted rope like the one shown in the figure above. Others first lay out their strings to outline an area and then mark one string three feet from the corner and the adjacent one four feet from the corner. Finally, using a tape measure they measure between the two marked points to see that they are five feet apart. If they are not, the strings will be adjusted so as to get the two marks five feet apart. Only then can the builder be sure the corners are square.

Four Corners

Based on a closed three-sided figure being known as a triangle, one might expect a four-sided one to be known as a quadrangle, or perhaps tetrangle, but, alas, it is neither. A closed four-sided figure is called a **quadrilateral,** and one is pictured below.

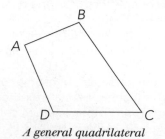

A general quadrilateral

"Quadrilateral" means four sides, so the name really does make sense. Quadrilateral *ABCD* is called a "general quadrilateral" because there are no special characteristics about it.

Quadrilateral *EFGH* is special.

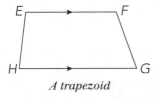

A trapezoid

Bases $\overline{EF}$ and $\overline{GH}$ are parallel to each other. A quadrilateral with one pair of opposite sides parallel is known as a **trapezoid.** If the legs of that trapezoid (the non-parallel sides) were equal in length, then it would be an isosceles trapezoid. What is true of any trapezoid is that two consecutive angles on each leg are supplementary. In the case of the trapezoid above, $m\angle E + m\angle H = 180°$, and $m\angle F + m\angle G = 180°$.

The next refinement in the family of quadrilaterals is achieved by making both pairs of opposite sides parallel.

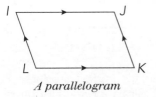

A parallelogram

IJKL is known as a parallelogram, defined as a quadrilateral with opposite sides parallel. ▱*IJKL* (the symbol means parallelogram) is marked as having two sets of parallel sides, but it also has some other features that may or may not be immediately apparent. Its opposite sides are equal in measure. So are its opposite angles ($m\angle I = m\angle K$; $m\angle L = m\angle J$). Any two consecutive angles are supplementary. What does that do for the sum of all of its angles?

Any quadrilateral can be split by a diagonal line segment into two triangles, so its angle measure must total that of two triangles, or 360°.

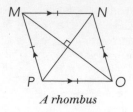

A rhombus

From the parallelogram, there are two directions in which we can go. We could make the parallelogram equilateral (all sides equal) in which case we would have a **rhombus**, such as ▱*MNOP* shown in the figure above. While the rhombus has the new quality of having all sides equal, it retains all of the properties of a parallelogram. It also adds one new property unique to the rhombus. The rhombus's diagonals ($\overline{MO}$ and $\overline{PQ}$) intersect to form right angles.

But we didn't have to go in that direction from the parallelogram. We could have chosen instead to add a right angle.

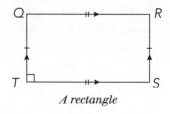

A rectangle

Adding a right angle to ▱*QRST* produces a possibly recognizable figure, the rectangle. Since a parallelogram's opposite angles are equal and its consecutive angles are supplementary, adding one right angle effectively adds four right angles. Think about it.

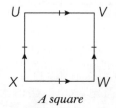

A square

Whether taking the rectangle and making it equilateral, or starting with a rhombus and adding a right angle, we end up with the most highly developed quadrilateral, the square. Do you think the diagonals of a square cross at right angles?* It's a rhombus, isn't it?

Pop Quiz

Answer each question in the most appropriate way.

1. A parking lot has four sides, two of which are parallel. What do we call the geometric figure with those properties?

2. A four-sided front lawn has one angle with a degree measure greater than 90°. What geometric figure does its shape describe?

3. If you were to add up the degree measures of the four corners of any four-sided closed structure, such as your garage, what would the total degree measure be?

4. Name two unique features that distinguish a rhombus from any other parallelogram.

Answers

1. trapezoid

2. quadrilateral (Question 2 was meant to trick you. A trapezoid has at least one angle greater than 90°, but it must have one set of parallel sides—a condition not mentioned in Question 2.)

3. 360°

4. It is equilateral and its diagonals cross at right angles. The same is true of a square, which, after all, is just a rhombus with a right angle.

Solid!

Many plane figures have solid equivalents—that is figures with the third dimension of height or depth. The simplest solid unit in the traditional system is the inch cube, or **cubic inch,** abbreviated cu. in., or in.^{3} In SI units it's the **cubic centimeter,** abbreviated cc, or cm^3.

---

* Yes, they do.

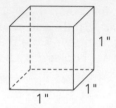

One cubic inch

Don't think of ice cubes as cubes. I've never seen one that was. The ice companies used to deliver blocks of ice that were cubes back before refrigerators were widely available, but that's ancient history. Today's ice cubes are half-moons, parallelepipeds (don't ask), or just about any shape imaginable—except cubes. To be a cube in the geometrical sense, the object must have straight edges that meet at 90°, pointy corners that form 90° angles, and identical measures of length, width, and height.

Count the number of **faces** on the inch cube above. Each inch square is a face. Now, how many **edges** does it have? An edge is the line segment between two faces. Finally, how many vertices (or corners) does it have? Your answers to the three questions should have been 6, 12, and 8, in that order. The faces are easy to spot and so are the corners. The edges are not that easy to count, which is why we've given you the following figure.

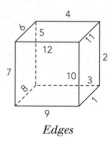

Edges

How many cubic inches would you expect to find in a cubic foot? If you're having trouble figuring that out, check out the following figure.

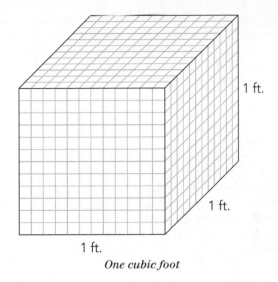

One cubic foot

If you figured out that there are 1728 cubic inches in 1 ft.^{3}, you did some very good thinking. One ft.^{3} is 12 in. long × 12 in. wide × 12 in. high. That's 12 × 12 × 12, or 1728 in.^{3} A unit raised to the exponent "^{3}" is read as that unit "cubed." In the SI system, a volume 10 cm long by 10 cm wide by 10 cm high is 1000 cm^3, or 1 liter. It's safe to conclude that when linear units are expressed as cubic measure, things get big quickly.

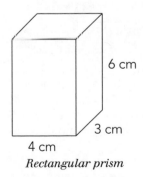

6 cm

3 cm

4 cm

Rectangular prism

The solid shown in the figure above is known as **a right rectangular prism**—rectangular because its base is shaped like a rectangle, and right because its base and sides form right angles with each other. You probably recognize it as a box. As a prism it takes up space. As a box, it has capacity to hold stuff. The question is "How much stuff?" For any right prism, the rule of thumb is to figure out the area of its base, and then multiply that by its height.

$$\text{Volume}_{\text{prism}} = \text{Area}_{\text{base}} \times \text{height}$$

The area of the base of this prism is 4 cm × 3 cm, or 12 cm². That times 6 cm gives it a volume of 72 cm³. Since in theory, the line segments that comprise the prism's edges take up no space, the solid occupies 72 cm³ or can hold 72 cm³ of stuff.

For any right prism, certain faces are rectangular along their edges. When a prism is a right rectangular one, all surfaces are rectangular. In the case of other ones, the non-rectangular faces are considered the bases,

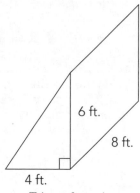

Triangular prism

In the above figure, we have a right triangular prism. Because the triangular side is the one not rectangular along all edges, we consider it to be the base. If you recall our study of area last chapter, then you recall that the area of a triangle is half that of a rectangle with the same base and height.

That means that the area of the base is ½ × 4 × 6, or 12 ft.² Multiply that by the height of the prism, 8 ft., to get 96 ft.³ And that's the volume of the right triangular prism in the figure above. Now for something almost completely different.

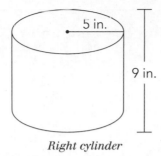

Right cylinder

The solid in the figure above is a **right circular cylinder.** Look at it closely. It is a right cylinder because the bases are perpendicular to the sides. Other than the round shape of its bases is it really any different from the prisms we've been looking at? It doesn't have rectangular sides, and yet the side could be unrolled to form one big rectangle. We'll look more closely at that in a short while. The volume of a right cylinder is found exactly the same way as that of a right prism. Find the area of the base, and then multiply that by the height. Do you remember how to find the area of a circle?

$$A_{circle} = \pi r^2$$
$$A_{circle} = \pi(5)^2$$
$$A_{circle} = 25\pi \text{ in.}^2$$

I'm sure you remembered that. There's no need to substitute for π, unless you actually need a numerical answer, which of course, in real life you would. Now, in order to get the volume or capacity of the cylinder, multiply what we just got by its height, 9 inches, to get 225π in.^{3}

It should be apparent that there are useful applications for the knowledge of how to find the volumes of prisms that are shaped like containers—especially rectangular and cylindrical ones. You can select the appropriate-sized container for packing a given volume of anything from candy to clothing to pots for storage. It is another story, however, when it comes to three other solids that are usually dealt with in geometry class.

Sphere, right pyramid, and right circular cone

Computing the Volume of a Sphere

The best way to compute the volume of a sphere would be to immerse the sphere in a full vessel of water which is sitting in a bowl. The amount of water that spills out into the bowl will equal the volume of the sphere—as Archimedes could have told you some 2500 years ago. You can then pour the overflow into a measuring cup to determine that volume. The same technique may be used with a pyramid or a cone. You will likely never encounter a pyramid unless you go to an Ancient Egypt exhibit, and the volume of a cone has no bearing on the amount of ice cream you can stuff into it.

Those figures are the **sphere,** the **pyramid,** and the **cone.** You are unlikely to encounter the sphere except as a ball, and there is certainly no need to compute its volume. If for some reason you wanted to, please see the sidebar on that subject.

Surface Area

Plane figures had a single surface, so finding their areas was a relatively straightforward thing. Solids, however, may have many surfaces. The surface area of a solid is the total of all the areas of its surfaces.

Finding the surface area of a cube is really rather straightforward. Look at the cube below.

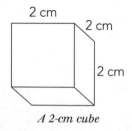

A 2-cm cube

What is the area of a single face of that cube? Did you figure out that 2 cm × 2 cm = 4 cm²? Now how many faces does the cube have? Six, so 6 × 4 cm² = 24 cm², and that's the cube's surface area.

Finding the surface area of the triangular prism in the following figure would take some work, and it's more than we need to do, but let's develop a plan.

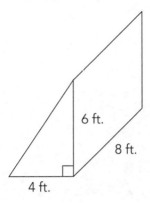

There are three rectangular faces and two triangular ones. The area of each triangular face is half its base times its height. That's $\frac{1}{2} \times 4 \times 6$, or 12 ft.^{2} There are two of those bases, so there are 24 ft.^{2} in the bases. One rectangular face is 4 ft. × 8 ft., one is 6 ft. by 8 ft. Their areas are 32 and 48 ft.^{2} respectively. To compute the area of the final side we'd need to use the Pythagorean Theorem to find the length of the triangles' hypotenuse. I'm not going to bother here, but if you'd like to, I'll provide the answer for you at the bottom of the page.*

Finally, consider the surface area of the right cylinder, such as the one shown below. It too has surface area, but it will take a bit of imagination to figure it out.

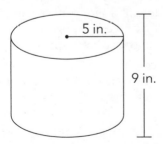

---

*$\sqrt{52} \approx 7.2$, so the area of the final side is 57.6 ft.^{2} That makes the total surface area 161.6 ft.^{2}

Consider each base of the right cylinder to be a circle, whose area can be computed by the formula $A = \pi r^2$. If you use your imagination, you might be able to picture the sides of the cylinder opened up and rolled out flat. What? No imagination? Then I guess you'll need to look at the following figure.

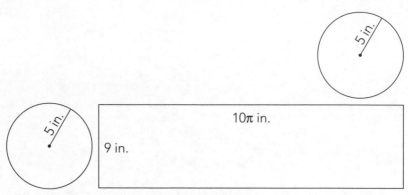

Peeling the cylinder

The height of the rectangle is the height of the cylinder. The length of the rectangle is what it was wrapped around—the circumference of the base, which is computed by $2 \times \pi \times r$. Make sense? Once again, I'll leave it to you to do the computations, but I will provide the answer (below).

Pop Quiz
Answer each question in the most appropriate way.

1. What is the volume of a block of ice you can make in a rectangular container that's 12 cm long, 8 cm wide, and 10 cm high? (Disregard the fact that water expands at 4° C.)

2. Assume that the container in Question 1 is closed and you intend to paint it. Paint is sold according to the amount of surface area it can cover. What is the surface area of the container in Question 1?

3. What is the volume of books that you could fit into a cube-shaped crate that's 12 in. long?

4. Assume the cube-shaped crate in Question 3 is closed and has solid sides, and you're going to paint it. Determine the surface area of that container.

Answers

1. 960 cm^3
2. 592 cm^2
3. 1728 in.^3
4. 864 in.^2

As for the surface area of the cylinder in the "Right cylinder" figure, as exploded in the "Peeling the cylinder" figure, each circular base has an area of $25\pi \text{ in.}^2$, for a total of $50\pi \text{ in.}^2$ The area of the rectangle is $90\pi \text{ in.}^2$ That's a total of $140\pi \text{ in.}^2$ If you use 3.14 as the approximation for π, you get 439.6 in.^2

9

Checkbook Math

Who ever heard of balancing a checkbook? Oh, I suppose a sea lion could do it on its nose, but seriously, I get it right most of the time, and I don't bounce more than one or two checks a year. Does this sound anything like you? I know I used to be like that. It used to be that checkbooks were only for writing checks. If you bounced a check occasionally, it cost you five or six bucks. So what? Then there's always overdraft checking. If you have overdraft checking and spend more than you have in the account, the bank'll fix it for you—for a fee. Ah, there's the rub! *For a fee.*

I used to think the bank was a place where I put money into a savings account every week, and every so often it paid me something called **interest.** That was to keep me interested in saving my money. What did I know? I was in second grade and keeping a bank savings account was a part of a New York City Public Schools' education curriculum.

What do you think a bank is? I'll tell you why a bank is in business. It's in business to make money. Banks take the money that savers put in and lend it out at high rates of interest, while thanking the savers by paying them a low rate of interest. That used to be enough for them, but not any more! Now, they're only too glad to charge you for anything they can get away with charging you for.

Furthermore, in addition to checks, a checkbook may also be used for keeping track of ATM transactions, some of which are made when buying stuff and some of which are made at an ATM cash-dispensing machine. Also, when you use an ATM machine that is not owned by your bank, you may incur charges (currently $0.50 to $3.00) made by your bank, and other charges of similar amount paid to the bank that owns the machine. Does any of this sound familiar? But wait, there's more. Banks now charge returned check fees of up to $40 per incident. The depositor of your bad check may be charged returned check fees up to the same amount, which (s)he is going to pass along to you. That means you could be paying as much as $6.00 to use an ATM machine and as much as $80 if you bounce a check. You could buy a check-writing calculator for that amount of money (on eBay) but it still wouldn't take care of the ATM charges. What's a body to do?

Register, but Don't Vote

Yes, you have to register to vote, but that's not the kind I'm talking about. Your checkbook has a register. It's the book of papers that aren't checks and have all those funny gray and white stripes and a bunch of columns with labels such as "Date, Number, Payment/Debt, Deposit/Credit" not necessarily in that order, and a final right-hand column with a "$" on its top. That's your check register. Ever notice it before? I'm sorry, I know I'm being flippant, but it has been estimated that 20% of checking account users never write in their registers, but rely upon memory and the once-a-month statements they get from their banks! My editor tells a story about Loretta, the bank manager, who chastised her for not using the register and, instead, recording her debit purchases and checks on the back of her checkbook. Then she'd check them off when they had cleared. You'd have thought she had committed the ultimate faux pas. My editor has been calling her check register "Loretta" ever since, in her bank manager's "honor."

RECONCILING YOUR CHECKBOOK

The first step in reconciling your checkbook is to use your register to record each and every transaction that you execute. The next step is to do the arithmetic, preferably as you go, but certainly before you turn to a new page in the register. That means adding in every deposit and subtracting every check written and every ATM transaction. You don't have to subtract to the nearest penny. Decide what level of accuracy you're happy with, and then stick with it. I've heard that rounding up or down to the nearer dollar usually works out. That means counting anything with a cent total of 50 or more as the next higher dollar, and counting everything with a cent total of 49 or less down as the lower whole dollar amount. I can't in good conscience recommend this method, since it is possible to end up with a small negative balance and have to pay the attendant penalties.

My wife, when she kept the checkbook, always kept it to the penny, and reconciled the balance each month when the bank statement came. She did that by using the back of the form that the bank sends. Keep in mind that you and your bank are never in agreement about how much money is in your account at any given time. That's because you've written a check or made a charge that the bank does not have a record of. Then there's that aggravating check that you gave to your nephew as a birthday present two months ago that's still sitting in the card because he hasn't yet needed to spend the money. That in itself is reason enough to keep track of what checks have cleared your bank and which have not yet been cashed. Not knowing that can give you a false sense of wealth that isn't really there.

Go find a bank statement. Oh, never mind, I'll use a copy of one of mine.

The back of the front page of your monthly checking account statement contains a chart that looks like the figure on the following page. Even if you never noticed it before, it's there. That's the place where you bring what your bank's version of your balance is in line with information you have, but they don't. That's what I mean by **reconciling.** The first column provides room for you to record any checks that you've written but that have not yet cleared the bank. The only way that you can figure that out is by going back at least two months and recording on a separate piece of paper the numbers of the checks that have cleared based on your current and your two previous statements. You don't need to know how much those checks were for. Since the bank has already paid them, those amounts won't be needed for your calculations.

CHECKS OUTSTANDING	
NUMBER	AMOUNT
TOTAL OUTSTANDING	

THIS FORM IS PROVIDED TO HELP YOU BALANCE YOUR CHECKBOOK WITH YOUR STATEMENT

1. If your account earned interest, enter the amount as it appears on the front of this statement in your checkbook.
2. Verify that checks are charged on statement for amount drawn.
3. Be sure that service charges (if any) or other authorized deductions shown on this statement have been deducted from your checkbook balance.
4. Verify that all deposits have been credited for same amount as on your records.
5. Be sure that all checks outstanding on previous statement have been included in this statement (otherwise, they are still outstanding).
6. Check off on the stubs or register in your checkbook each of the checks paid by us.
7. Make a list of the numbers and amounts of those checks still outstanding in the space provided at the left.

8.	ENTER FINAL BALANCE AS PER STATEMENT		
9.	ADD ANY DEPOSITS NOT CREDITED		
10.	TOTAL		
11.	SUBTRACT ATM WITHDRAWALS + AUTOMATIC PAYMENTS (NOT SHOWN ON STATEMENT)		
12.	SUBTRACT CHECKS OUTSTANDING AND OTHER DEBIT TRANSACTIONS		
13.	BALANCE SHOULD AGREE WITH YOUR CHECKBOOK		

CARRY OVER →

Watch for Weird Things on Your Statements

This book is not meant to be one of anecdotes, but this happened to me less than two weeks ago. It is a good lesson for why you need to watch your checking statements closely. I was noticing that for the last several months my checking account always seemed to be coming up a few bucks short. Somewhere between $12 and $20 every month or two, and I attributed it to either my wife or I having been careless about recording something that one of us had charged to our ATM card. Since banking has gone on line, I look in on my checking account every week or so, sometimes as often as two or three times in a week. One day I noticed a charge of $14.95 noted as "Debit Memo." From experience I knew that it would take a day or two for the actual source of the charge to show up, but I also knew that I hadn't spent $14.95 on anything, and my wife claimed not to have either. The following day the charge showed up, attributed to some internet address that I'd never heard of. I went to the address and saw that they were offering something for a ten-day

By the way, you can't just reconcile your account at any time. The best time to start is right after you've received your bank's statement, although if you go on line, you don't have to wait until then. You can also find out your balance and current activity (check clearing, etc.) over the phone. The day your monthly statement arrives, all the latest information is at your fingertips. The reason for going back a couple of months is just in case there's that aggravating nephew out there with the uncashed check.

Once you have the numbers of all your checks that the bank has cleared, go to your check register and find the numbers and amounts of all the checks you have written that have not cleared. Enter them one line at a time in the "Outstanding Checks" column on the back of the bank statement, while bearing in mind that "outstanding" in this context does not mean "way cool." Total those amounts at the bottom of the column (that's why you bought the calculator in the first place), and move them to the right into the space labeled "12."

If your checking account earned interest, enter the amount into your checkbook as a deposit. Next look and see if you have made any free trial, which then charged $14.95 per month unless it was cancelled. That amount rang a bell, and so did the service. I had accepted the free trial awhile back, cancelled the service within a day as being of no real use to me, and had forgotten about it.

When I called them to complain, I was told that they were sorry I wanted to discontinue my service after having been signed up for a year. I was incredulous and asked to speak with a supervisor, to whom I explained the whole situation. He said emails discontinuing the service were not recognized since they went to an automatic responder, and one could only cancel by phone. Only after I threatened to sue did he agree to refund me five months of fees for the service. Despite my argument that their records should show that I never used the service, he said they were like cable television. Once it was installed, the cable company didn't care whether or not you watched. Of the $165 I paid in for a service I didn't want I got $75 back, but my checking account shouldn't need those mysterious adjustments any more. My wife thinks she should be allowed to spend the equivalent of what I lost on eBay.

deposits to your account that were not credited on your statement. Hey, nobody said it was going to be easy. The total of the interest and those deposits should be placed in the space labeled "9."

In space 8 write the total from your statement (what the bank believes to be your balance). Next add the amounts in spaces 8 and 9 and write the total in space 10. Then find any ATM withdrawals or automatic payments that you've scheduled but which do not appear on your statement. Add them up and put the total in the top part of space 11. Draw a line under them and subtract from the amount in 10. Finally, subtract the amount in 12 from what you just got in 11. That is your reconciled balance—the amount that should be the same as that on the bottom right of your checkbook register.

Are the numbers close? If the reconciled balance is less than what's in your checkbook you're okay. My wife claims her reconciled balance always was perfect to the penny. She also claims to be able to walk on water. I've never asked her to prove either claim. If the reconciled balance is higher than what's in your checkbook, you'd better find a mistake the bank made, or rush to deposit the difference.

An Alternate Strategy

There is no excuse for not entering your transactions and adjusting your running balance unless you're a glutton for punishment or wealthy enough to afford an accountant to do it for you. For that matter, you could use a spreadsheet such as Lotus or Excel, but if you're too lazy to write amounts in your check register, what are the odds that you're going to keep a spreadsheet or even a checkbook program up to date? Of course there are degrees of accuracy. In an earlier section I rejected the idea of rounding up and rounding down to keep track of your balance, since there's no guarantee that you can't overdraw, but what if you rounded in only one direction–up?

You keep the dollars and cents in the "Payment/Withdrawal" column but you round up in the running totals (right-hand) column. So when you spend $10.75, you enter $11; when you spend $21.30, you enter $22.00. For every amount you spend that is not in even dollars, you round the entry up to the next dollar. That eliminates the need to subtract pennies ever again. Pretty soon you'll have built a bit of a cushion against future errors.

But why stop at the next dollar. Do you have trouble saving? You might try rounding up to the nearest $5 or even the nearest $10. In short order you'll have built up some savings. Then, once a month, reconcile your checking account and switch your savings into a high-interest savings account, such as the internet one that's always advertising on TV. You'll never pay extra bank fees, and you'll save for that proverbial rainy day almost painlessly. I don't know about you, but I like it!

Pop Quiz
For the sake of argument, round up each entry to the next dollar

1. Mary has a checking balance of $511. She pays $42.39 with her ATM card at the gas pump, writes a $79.36 check to the electric company, an $82.11 check for groceries, and withdraws $60 in cash from an ATM belonging to her own bank. What will her new balance be?

2. Jason's checking account has a balance of $723. He writes a $56.45 to the cable company, an $89.36 check to the gas company, and a $47.85 check to his credit card company. Next he stops by the supermarket, where he puts a $57.93 total on his ATM card and takes $20 in cash back. Having spent that at the stationery store, he uses an ATM machine that charges $1.50 to do a fast cash withdrawal of $60. He knows that his bank will also charge him $1.50 for using a different bank's ATM. What will his new balance be?

3. What is the name given to bringing your checkbook into agreement with your bank's statement?

4. When is the best time to do the above?

Answers

1. $245
2. $386
3. Reconciling (No, it's not "Loretta-ing" ☺)
4. The best time to do it is right after you've received your bank's statement.

10

Credit Card Math

Ah, credit cards! What a wonderful convenience. You don't have to carry cash on your vacation, so you minimize the likelihood of being robbed (more on that in a moment). If your purchase doesn't prove to be as advertised, you can notify the credit card company and have them withhold payment. And, if you pay off your total balance within the specified grace period, you don't pay any interest.

Take an Interest in Interest!

Under the practice of risk-based pricing, the interest rate on a loan—and a credit card is a form of loan—should reflect the risk to the lender. This may also be interpreted as allowing the lender to grant loans to a wider group of customers with varied credit histories, charging the lower interest rates to those with the better histories and higher rates to those whose credit histories are spottier. The loan's interest rate should not cause the borrower who always pays on time to have to subsidize those customers who **default** (don't pay). With most loans, the risk premium (amount charged for extra risk) is set at the time of borrowing, but since with a credit card the amount borrowed changes with every new purchase the cardholder makes, the risk may change later. We'll look at this more a little later in this chapter, in the section entitled "Universal Default."

Now let's get down to reality! Credit cards carry a high rate of interest–even the low interest ones. Furthermore, most of what you are charged each month on outstanding balances is interest. Paying the minimum amount on a credit card means it can take you thirty-five to forty years to pay off the balance, and most of what you pay during that time will be interest. No wonder banks love them. You used to need to have an excellent credit rating/record to qualify for a credit card. Now banks are throwing them at college students, people living barely above the poverty level, and those already deeply in debt, telling them how credit-worthy they are. Why is this happening? Because, for the issuing institution, credit cards are gold mines. Where else can a bank put its money and expect a 14 to 30% return in perpetuity? You know the answer to that question.

CALCULATING PERIODIC FINANCE CHARGES

Have you ever read the back of your credit card statement–assuming you have one or more cards (almost everybody does). Let me summarize the back of one of mine, while keeping the institution's name— a nationally recognized bank—to myself:

"The grace period for purchases is at least 20 days [it used to be 30].* To avoid periodic finance charges on purchases that appear on this statement, you must have paid the new balance on the last statement by that statement's payment due date and also pay the new balance on this statement by this statement's payment due date. If you made a balance transfer, you may not be able to avoid periodic finance charges on new purchases as described in the balance transfer offer. [At least they warn you, albeit in small light print.] Your annual percentage rates (APR) and periodic rates may vary. There is no grace period for cash advances.

"Periodic finance charges are calculated separately for each balance subject to different terms, for example, new purchases may be billed at a different rate from that on old purchases [and cash advances are sure to carry the highest interest rate]. Charges include purchases, balance transfers, cash advances, transaction fees, other fees, and any minimum finance charge.

"For each balance, the daily balance is multiplied by the periodic rate [determined by dividing your annual rate by 365 and rounding it to 5 decimal places]."

Without going into all the details here, what you're being told is that your interest rate is compounded on a daily basis, and, as we've discussed earlier, compounding increases interest fast. Do take the time to read the back of one of your statements. It is very informative and is something the credit card institution doesn't want you to know, but tells you because the law requires it of them. To make it as inconspicuous as possible, it's placed on the back of your statement and is printed in a light-colored ink in the hope that you'll never notice it.

---

*Items in brackets are my observations!

What Happens If You Just Pay the Minimum?

Beginning in 2006, most U.S. credit card companies regulated by the Office of the Comptroller of the Currency have been required to increase customers' minimum payments to cover at least the interest and late fees from the prior statement plus 1% of the outstanding balance.

The computations below apply to credit cards carrying the Visa or Mastercard logos. Discover is a bit different, and American Express has two different revolving credit cards, and one that must be paid in full at the end of the period for which the items were charged to it.

Now let's look at the front of my statement, or yours. Let's assume a balance of $5000 and an APR of 16.5%, which is pretty nominal in today's mix. An APR of 16.5% translates to a decimal value of 0.165. (If you're not sure how to convert percents to decimals, go check back to Chapter 6, Using Percents.) Divide that by 365 to get the daily (periodic) rate.

$$\frac{0.165}{365} = 0.000452 \text{ (or } 0.452\%)$$

So the following table shows what you'll owe after one 30-day month of daily compounding. In each case, the previous day's total has been compounded. As it so happens, however, at this rate, peculiarly, after the third day the rate stays the same. The total owed, however, does not:

30 Days of Compounding		
Day #	*Interest*	*Total*
1	2.260271774	$5,002.26
2	2.26129354	$5,004.52
3	2.261294002	$5,006.78
4	2.261294002	$5,009.04
5	2.261294002	$5,011.31
6	2.261294002	$5,013.57
7	2.261294002	$5,015.83

Day #	Interest	Total
8	2.261294002	$5,018.09
9	2.261294002	$5,020.35
10	2.261294002	$5,022.61
11	2.261294002	$5,024.87
12	2.261294002	$5,027.13
13	2.261294002	$5,029.40
14	2.261294002	$5,031.66
15	2.261294002	$5,033.92
16	2.261294002	$5,036.18
17	2.261294002	$5,038.44
18	2.261294002	$5,040.70
19	2.261294002	$5,042.96
20	2.261294002	$5,045.22
21	2.261294002	$5,047.49
22	2.261294002	$5,049.75
23	2.261294002	$5,052.01
24	2.261294002	$5,054.27
25	2.261294002	$5,056.53
26	2.261294002	$5,058.79
27	2.261294002	$5,061.05
28	2.261294002	$5,063.32
29	2.261294002	$5,065.58
30	2.261294002	$5,067.84

Notice that almost $68 in interest has accumulated in just thirty days. The amount you'll be expected to pay is at least 1% of the principal, or $50, plus the interest of $67.84. That's $117.84. Notice that more than half of what you're paying back is interest. At that rate, it would take you years and years and years to pay it off. Figure if you're paying 1% of the principal each month, it's going to take quite a bit of time to pay off the entire principal, and, in fact, you're going to end up paying more interest than the $5000 worth of merchandise you charged.

By the way, that 1% of the principal rule, as noted above, is a relatively new thing. Under the old law practically everything you paid was interest, and you could take forty years or more to actually pay off the card. The conclusion you should be reaching here is that if you're paying off a credit card, don't make just minimum payments. The less you pay each time, the more interest you'll pay over the length of the pay-off time.

OTHER REASONS TO PAY CASH

In case you haven't figured it out yet, credit card debt has the potential to be a nightmare scenario. Did that video camera look too juicy to pass up at the sale price of $500? Well, if you put it on your credit card and didn't pay it off during the grace period (however long yours happens to be), figure its price just increased by one half. It doesn't appear quite such a bargain at $750, does it? That is the best way to think of any purchase you're planning to make with a credit card. If you could wait to buy it if its price were 50% more than it is, then wait. There are advantages to paying cash.

Credit card interest rates vary from time to time and are pegged to the prime interest rate according to a formula determined by the Federal Reserve Board. That board charges a certain rate to the bank to lend it money, and the bank then turns around and charges you more. Furthermore, your balance hops around from month to month depending on your interest rate, special fees, late charges, if any, and your purchases. Figuring out your interest payment at any given time is in reality an amazingly complicated job and is best left to a special computer program.

Then there's also the matter of late fees. If you miss the due date by as much as a few minutes, you'll be charged a late fee and it's no nominal amount. In general, late fees can vary from $30 to $40. Furthermore, paying your bill late can also send your moderate to high interest rate skyrocketing.

Universal Default

A rather nasty practice that is being used by more and more credit card issuers is one known as **universal default.** It is common practice in the credit card industry for the lender to change the terms of a credit card from time to time, and then notify the cardholders. The default terms are the least desirable terms (highest interest rate), and are generally charged to those customers who have defaulted (failed to pay in a timely manner) on their loan. Universal default is the practice of Lender A's raising a customer's terms to the default level upon the news that the customer has defaulted with another lender, say Lender B, even though the account with Lender A is still in good standing. The growing use of universal default is one of the most controversial practices in the credit card industry.

Secured Credit Cards

One more credit card trap awaits the unsuspecting consumer, and that is the secured credit card. A secured credit card is a type of credit card that is secured by a deposit account owned by the cardholder. As a rule, the cardholder has to deposit somewhere between 100% and 200% of the total amount of credit being sought. So, if you put down $1000, you'll be given credit in the range of $500 to $1000. In some cases, credit card companies may offer incentives even on their secured cards. In those instances, the deposit required may be substantially less than the requested credit limit, as low as 10% of the desired limit. This deposit is held in a "special" savings account. What is so special about it is that you can make additional deposits, but you can't make withdrawals—at least not while you have outstanding debt on the secured card.

The holder of a secured credit card is still expected to make regular payments, as with a regular credit card, but should he or she default on a payment, the card issuer has the option of recovering the cost of the purchases paid to the merchants by withdrawing it from the cardholder's savings account.

Although the deposit is in the hands of the credit card issuer as security in the event of default by the consumer, the security will not be forfeited simply for missing one or two payments. That would be too easy and would let the cardholder off the hook. Usually the deposit is only claimed as an offset when the account is closed, either at the customer's request or due to severe delinquency (150 to 180 days). This means that an account that's less than 150 days overdue will continue to accrue interest and fees, and could result in a balance much higher than the actual credit limit on the card. In fact, the total debt may far exceed the original deposit and the cardholder not only forfeits the deposit but may be left with additional debt to pay off. Pretty good deal for the card issuer, don't you think?

Pop Quiz

1. Reese has a $5000 balance on a credit card that has an annual rate of interest of 13%. Given the new government requirements mentioned in the beginning of the chapter, how much should she expect the minimum payment should be on her next bill?

2. Alex missed making his February payment at 16% interest on his $8000 credit card balance. Now, he must pay a $30 late penalty and his interest has been raised to the penalty level of 32%.

   a) What would it have cost him if he had payed on time?

   b) How much greater will his February payment be than if he had made it on time?

3. Ian has three credit cards. Card A has a balance of $6500 at 12% annual interest; Card B's balance is $4800 with an annual interest rate of 17%; Card C's balance is $5300 at an annual interest rate of 15.5%. Ian is late two times on his payment on Card A, thereby triggering universal default to kick in, with all three cards going to 31% interest. Disregarding any late fees, how much more will next month's payment cost him than it would have if he had always been on time?

Answers

1. $104.15

   13% ÷ 12 months = 1.083%, or .01083

   $5000 × .01083 = $54.15

   To that, add 1% of the principal, or $50.

   $50 + $54.15 = $104.15

2. a) $186.64. On time, his payment would have been 16% ÷ 12 months = 1.333% or .01333. $8000 × .01333 = $106.64. To that, add 12% of $8000, or $80 for a total of $186.64.

b) $136.64. For not having paid on time, his 16% interest has been doubled to 32%. That's an increase of $106.64 and he is charged a $30 late fee for a total penalty of $136.64 for February alone. If the increased interest rate remains, he will pay twice as much interest for each succeeding month.

3. $268.89. Card A increases by (31% – 12%) = 19% annually. That's a monthly increase of 19 ÷ 12 = 1.583%.

.01583 × $6500 = $120.25.

Card B increases by (31% – 17%) = 14% annually. That's a monthly increase of 14 ÷ 12 = 1.167%. .0167 × $4800 = $80.16.

Card C increases by (31% – 15.5%) = 15.5% annually. That's a monthly increase of 15.5% ÷ 12 = 1.292%.

.01292 × $5300 = $68.48.

Now we add the three together:

$120.25 + $80.16 + $68.48 = $268.89

11

Mortgages

A mortgage is probably the largest investment you'll ever make on anything. It's the loan you take out in order to buy a house. There are basically two types of mortgage, fixed rate and adjustable. The amount of money that you borrow is known as the **principal.** It is on that amount that the **interest** (what it costs you to borrow that money) is computed. Whether fixed or adjustable, the computation is pretty straightforward, but I'll provide you with a table to make it easier.

Adjustable rate mortgages, or **ARM**s, are usually cheaper (that is, have a lower interest rate) than fixed rate mortgages. They are pinned to, and somewhat more than, what the Federal Reserve Bank charges the commercial or savings bank for funds. Such mortgages are usually convertible to fixed rate after a period of one year, and before five or ten years have elapsed. When the prime interest rate goes up, the ARM's interest rate goes up; when the prime rate goes down, the ARM does the same. You'll have to admit that this is a pretty good arrangement—for the lending institution. The up and down adjustments (which in my experience were almost always up) are usually made at fixed intervals such as quarterly, semiannually, or annually. When you choose to lock into a rate, you pay a nominal fee ($100 to $200) to do so. That allows you to lock into the interest that prevails at the time for fixed-rate loans.

There is usually a maximum amount of variation that an ARM can go through. In the case of the one I had, it was 6 points in either direction. I took my ARM out at 8.5%, and it managed over the course of four years to increase to 14%. That highlights the danger of an ARM. What appears to be a good deal at the start can end up going to a point where you can't afford to make the payments. When you opt for an ARM you are literally betting the farm on that mortgage's staying affordable. Just for the record, I locked in my ARM when it dropped to 10%. Of course, rates continued to drop, to my chagrin, and I finally refinanced at 6.375% (thanks for asking).

The $100,000 Question

Now for that table. Be aware that none of these numbers are exact amounts, but have been rounded to the nearest dollar. The 5% figure, for example, is actually $536.82, but why bother to worry about pennies when we're dealing with hundreds or thousands of dollars?

Thirty-Year Rates for $100,000 Mortgage

Interest Rate	Monthly Payment
5%	$537
6%	$600
7%	$665
8%	$733
9%	$805
10%	$878
11%	$952
12%	$1029
13%	$1106

This table provides the computed monthly payments on mortgages of $100,000—a very nominal amount these days—at different percentage rates ranging from 5% to 13%. It is unlikely that mortgage rates will ever go below 5%—they haven't in my lifetime—but if they do, or (heaven forbid) exceed 13%, just go to your favorite internet search engine and type in "mortgage calculator." One is available free from a multitude of sources. In fact, that's where I got the numbers in both tables in this chapter.

Let's assume for a moment that you've shopped around for a mortgage—and it definitely pays to shop for the best rate you can get—and have settled on a thirty-year one for exactly $100,000 at 7% interest. Theoretically, your monthly payment should be $665. I say "theoretically," because that doesn't take into account the fact that your property taxes and homeowner insurance will in all likelihood be divided up for the year into twelve parts, and both of those will be added to your payment. So if your insurance is $600 per year and your property (and school) taxes are $3000, $1/12$ of that is $300, which will bring your actual monthly payments to $965. But never mind about that right now. After all, you haven't even closed yet. By the way, closing is going to cost you also—and I can't tell you how much, because terms vary considerably by home price, lender, lawyer's fees, and more. Closing costs represent only a small part of the expense associated with buying a house.

At the closing, you are going to be asked to sign a note, and that note will be for the full amount being borrowed. Let's see now. Thirty years is 360 months. At $665 per month, $360 \times 665 = \$239,400$. Surprised? That's why at the closing they ask you to take a seat long before they ever show you the note. You'll pay almost $140,000 in interest on the $100,000 borrowed.

But wait. What if your mortgage were taken out at 9%?

$$805 \times 360 = \$289,800$$

That's an additional $50,000 for the 2% increase in mortgage interest rate. That's a lot of money! Of course, as I've already alluded to, mortgage interest rates don't vary by whole percentage points. They can be offered as something and a quarter or something and a half, not to mention the $6 3/8$% of my own mortgage rate. I'm not going to tell you how to compute those rates exactly, but I'll show you how to get very (within a few dollars) close.

Consider borrowing $100,000 at a rate of $7 1/2$%. At 7% the monthly payment would be $665; at 8% it would be $773. You can bet that $7 1/2$% is going to be somewhere in between. The difference between the two rates is $773 - 665 = \$108$. Half of that is $54. Add that to the lower amount, and you'll get $665 + 54 = \$719$ as the monthly rate. That'll be pretty darned close to $7 1/2$%.

All right. This all works very well for $100,000 loans, but what if you want to borrow $89,000 or $127,000. Then what? Good point, but before I can answer that, you need to recall ratio and proportion. We studied that back in Chapter 8 when we studied similar triangles. If you need refreshing on that, go back to page 109 and take a look.

An $89,000 Mortgage at 8%

To figure the rate of a thirty-year mortgage at 8%, we're going to need to make a proportion. We'll make the proportion easier to work with by using thousands of dollars. We're looking to equate the ratio 89,000:100,000 with the ratio looked-for amount:733 (the monthly rate for 100,000 at 8%, from the previous table), or

$$\frac{89}{100} = \frac{\text{looked–for amount}}{733}$$

If you're wondering why we didn't need all the zeroes, it's because they would have cancelled themselves out once we had placed the 89,000 over the 100,000. Next, you'll recall, since the product of the means equals the product of the extremes, multiply as follows:

$$100 \times (\text{looked-for amount}) = 89 \times 733$$
$$100 \times (\text{looked-for amount}) = 65,237$$

Right now, the looked-for amount is being multiplied by 100. To get it by itself, we need to divide both sides of the equation by 100 (since what you do to one side of an equation doesn't change its value if you do the same thing to the other).

$$\frac{100 \times (\text{looked–for amount})}{100} = \frac{65,237}{100}$$
$$\text{Looked-for amount} = 652.37 = \$652.37$$

There we have it. The monthly payment for a thirty-year mortgage of $89,000 is about $652.

What's that? You say that looked suspiciously like algebra? Well, I promise I won't tell anyone if you don't.

A $159,000 Mortgage at 7%

To figure the rate of a thirty-year mortgage at 7%, again we're going to need to make a proportion. Actually, the procedure will be almost identical to the one we used to find the 8% loan above, but I'm going to indulge myself and use the letter x to stand for the "looked-for amount." If that scares you, just read the x as "the looked-for amount." It's only a placeholder, and takes up a lot less space, not to mention typing. We're looking to equate the ratio 159,000:100,000 with the ratio x:665, (the monthly rate for 100,000 at 7%, from the preceding table), or

$$\frac{159}{100} = \frac{x}{665}$$

Next, multiply the means and the extremes.

$100x = (159)(665)$ That's "times" on both sides.

$100x = 105,735$

Finally, divide both sides by 100 to get the "looked-for amount" by itself.

$x = 1057.35$, or $1057.35

The monthly amount would be about $1057. Would you like to know how big a mortgage you'd have to take out and the part of the repayment amount that will be interest? Just use the rounded figure $1057 and multiply by the number of months in 30 years.

$1057 \times 360 = \$380,520$ for the total of the note.

$380,520 - 159,000 = \$221,520$ in interest.

That, in case you couldn't tell, is a huge amount of interest on a principal amount of $159,000.

Alternative Terms

Thirty-year term mortgages became popular when interest rates were soaring in the 1970s. They were the only way the vast majority of middle-class home buyers could afford to finance their purchase, and they remain the most popular instrument of home buying today, but they have never been the only one. Mortgages have generally been available in fifteen-year and twenty-year terms. The following table shows some monthly rates side-by-side for a $100,000 loan.

Monthly Rates for $100,000 Mortgage Over Different Terms			
Percentage	**15 Years**	**20 Years**	**30 Years**
5%	$791	$660	$537
6%	$843	$716	$600
7%	$898	$775	$665
8%	$956	$836	$733

Percentage	15 Years	20 Years	30 Years
9%	$1014	$900	$805
10%	$1075	$965	$878
11%	$1137	$1032	$952
12%	$1200	$1101	$1029
13%	$1265	$1172	$1106

Look at the 5% rate for a fifteen-year mortgage and compare it to the same rate for thirty years. Why do you suppose they are so close to one another? In fact, they're just slightly more than $150 apart, even though one term is twice the length of the other. To answer that question, let's figure out the cost of repaying the note. The thirty-year mortgage requires 360 payments of $537. That's a total of $193,320. The fifteen-year mortgage requires 180 payments of $791 for a total of $142,380. That's $93,320 interest on the longer term versus $42,380 on the shorter one—a difference of about $50,000 spread over the mortgage's term. That's why the monthly repayment amounts are so close to one another.

Closer still is the twenty-year mortgage's monthly repayment amount. Twenty years is 240 months, so the total amount comes to $240 \times 660 = \$158,400$. That's $58,400 in interest, or about $16,000 more than the fifteen-year mortgage. That still seems to be quite a bargain when compared to the thirty-year one.

I'll leave it to you to eventually figure out the difference in interest payments at 13%. You'll probably be shocked.

Faster Payback

Most banks these days also offer a bi-weekly repayment plan, where you pay half a mortgage payment every two weeks. Normally you would pay your mortgage once a month, or twelve times per year. With the bi-weekly payback plan you are paying half the monthly amount twenty-six times per year, or twenty-six halves, which is the same as thirteen full payments per year. Without getting into the calculations, since you are repaying your mortgage more quickly with this plan, you will end up paying your thirty-year mortgage in about twenty-three years, and consequently pay tens of thousands of dollars less in interest. What you're not being told, of course, is that you have to be able to afford to pay an extra month's mortgage each year. Some people can, and some can't.

Whether or not you can afford to pay that extra full month's mortgage or not, you should still consider paying extra principal whenever you can. Whether it's $20 per month or $200 per month, or just an irregular amount at irregular intervals, almost every mortgage contract I've ever seen permits the lender to pay extra money specified as funds to be applied to the principal. The sooner you have succeeded in paying back the principal, whether it is months early or years sooner than the schedule, the more the total amount of interest you will pay will be reduced.

When to Refinance

There is no simple formula for when to refinance a mortgage, nor how long it will take for you to break even on a refinance. That depends on many factors, including your current interest rate, the new interest rate, what the closing costs are, how much money you are looking for, and how long you plan to stay in your home.

There are three basic reasons for refinancing. First is to take advantage of falling interest rates. Usually, if you can refinance at a rate 2% or more below your current interest rate it pays to do so. More about this in a moment.

Another reason to refinance is to get money out of the equity in your home in order to pursue some project, such as a major renovation. Refinancing your home in order to buy a car, however, is a very poor idea, since you will be repaying the money you borrowed for the car for many years after the car's value has depreciated to nothing.

The third reason for refinancing is to pay off high-interest and debilitating credit card accounts. Mortgage interest is tax deductible; credit card interest is not.

Beware of the trap. If you are increasing your mortgage obligation to get out of credit card debt, cut up those credit cards. I personally know of people who have run up over $30,000 in credit card debt, got out from under it by refinancing their home, and then, with their vastly improved standing with Visa and Mastercard, ran up another $32,000 in credit card debt. Bankruptcy is probably on their horizon.

As you pay down your mortgage, your house accrues **equity.** That's value. The greater the difference between the market value of your house and the amount of principal you still owe, the greater your equity. Be careful about this, too. House market values can go up and they can go down. Throughout the 1980s through half of 2006, they went up, up, up, but in 2006, they peaked and began to go down. Some mortgage bankers will allow you to borrow up to 80% of the market value of your house. Some will allow up to 125% of its value. Avoid being seduced by temptation. Also, as before with a first mortgage, shop for the best deal. Don't take the first one that comes along. As a general rule, those lenders that advertise on television or send you unsolicited offers in the mail are **not** the best bet. Use the internet to find out the experiences of other borrowers who have dealt with the lender(s) you are interested in. There's a lot of valuable information on line, but *know who your information provider is.*

Finally, I promised you more information on calculating the best deal(s) on when it pays to refinance. As with mortgage calculators, there are great on line refinance calculators. One worth looking at is www.dinkytown.net/java/MortgageRefinance.html. With it, you can calculate breakeven point, closing costs, monthly payment breakdown, and more. Take advantage of this and other on line resources.

Pop Quiz

Use the rates in the previous figure to compute your answers. Express your answer in whole dollars.

1. What is the difference in the interest paid over the life of a $100,000 mortgage at 13% interest for thirty years versus fifteen years?

2. What would be the cost of the mortgage in Question 1 for a term of twenty years?

3. What would be the monthly cost of a $228,000 mortgage for a period of thirty years at an interest rate of 9%?

4. How much more or less would the mortgage in Question 3 cost per month if it were taken out for a twenty-year term at the same rate?

Answers

1. $170,460
2. $281,280
3. $1835
4. $217 more

12

Insurance Math

It is a basic tenet of the gambling casino industry that the house always wins. In that case, why does anybody ever go to a casino? Well, the fact of the matter is that "the house always wins" is a long-term proposition not to be taken literally. It means that given a large number of gamblers over a reasonable period of time, there will be more individual losers than winners, and so the house will come out ahead. The same rules that govern the casino industry apply to the insurance industry, both being governed by the branch of mathematics known as **probability.**

Probability

The probability of something occurring is a number based on how many ways the desired outcome can occur vs. how many total outcomes are possible. It is expressed as a fraction, determined by:

$$\frac{\text{desired outcome(s)}}{\text{possible outcomes}}$$

A probability of 1 is known as **certainty;** a probability of 0 is known as **impossibility.** Most events have a probability somewhere between those two extremes. When you toss a coin, there are two possible outcomes, a head or a tail. Suppose you call heads. When tossing a single coin, you have one chance out of two to toss a head, or a probability of $\frac{1}{2}$. The probability of tossing a tail is also $\frac{1}{2}$. Notice the sum of the probabilities of tossing a head or a tail is certainty, since $\frac{1}{2} + \frac{1}{2} = 1$.

What about a single die (one of two dice)? What is the probability of tossing a single die and having it come up 6? Well, a die has six sides, only one of which has six dots on it, so the probability of rolling a 6 is 1 in 6, or $\frac{1}{6}$. The probability of rolling a non-six is $\frac{5}{6}$. And $\frac{5}{6}$ is 5 times as great as $\frac{1}{6}$. That makes the **odds** 5 to 1 against rolling a 6, or 1:5 for rolling it. That's right, odds are a ratio comparing two probabilities.

So far this has been pretty simple. Let's make it just a little more complicated, and a little more practical at the same time. Consider rolling two dice. How many possible ways can they fall? Well, there are really two answers to that question. The first, and most geometrically correct, answer is that since each die can come up 6 different ways, there are 6×6 possible ways, or 36 ways that they might come up. For all practical purposes, that makes the probability of throwing boxcars (a 12) $\frac{1}{36}$. The probability of throwing snake eyes (2) is the same, but then things change. What's the probability of two dice coming up to total three? Here's where order comes into the picture. We have to consider the two different dice as individuals even though they are being thrown together in one toss. Let's call the left die "L" and the right die "R." We can throw a 3 if L = 1 and R = 2, or L = 2 and R = 1. Is there any other way to throw a 3 using two dice? So what's the probability of throwing a 3? There are two favorable outcomes over thirty-six possible outcomes. That's $\frac{2}{36}$, which simplifies to $\frac{1}{18}$.

A 4 can be thrown as 1 + 3, 3 + 1, and 2 + 2. That makes the probability of throwing a 4 $\frac{3}{36}$, or $\frac{1}{12}$. You can work out the other probabilities, but let's try one more together. How many ways can you make a 7? There's 1 + 6, 6 + 1, 2 + 5, 5 + 2, 3 + 4, and 4 + 3. That's a probability of $\frac{6}{36}$, or, simplified, $\frac{1}{6}$, of rolling a 7. Bear that in mind, the next time you're tempted by the craps table.

Term Life

"But what does probability have to do with the price of beans?" you might well be asking. The answer, I'm afraid, is everything. The insurance company is betting that you'll live longer than the insurance that you're buying. They make that bet based on probability. The Period Life Table (see following page) is a portion of an actuarial table for period (also known as term) life insurance.

Period Life Table, 2002

Exact Age	Male			Female		
	Death probability a	Number of lives b	Life expectancy	Death probability a	Number of lives b	Life expectancy
30	0.001389	97,094	45.98	0.000628	98,424	50.58
31	0.001428	96,959	45.04	0.000673	98,362	49.61
32	0.001484	96,821	44.1	0.000727	98,296	48.65
33	0.001561	96,677	43.17	0.000793	98,224	47.68
34	0.001657	96,526	42.24	0.000869	98,146	46.72
35	0.00177	96,366	41.31	0.000953	98,061	45.76
36	0.001897	96,196	40.38	0.001045	97,968	44.8
37	0.002043	96,013	39.45	0.001147	97,865	43.85
38	0.002207	95,817	38.53	0.001259	97,753	42.9
39	0.002389	95,606	37.62	0.001381	97,630	41.95
40	0.002589	95,377	36.71	0.001514	97,495	41.01
41	0.002808	95,130	35.8	0.001655	97,347	40.07
42	0.003047	94,863	34.9	0.0018	97,186	39.14
43	0.003306	94,574	34	0.001946	97,011	38.21
44	0.003585	94,262	33.12	0.002097	96,823	37.28
45	0.003891	93,924	32.23	0.002264	96,620	36.36
46	0.004218	93,558	31.36	0.002446	96,401	35.44
47	0.004554	93,164	30.49	0.002639	96,165	34.52
48	0.004895	92,739	29.63	0.002816	95,912	33.61
49	0.005249	92,285	28.77	0.00301	95,642	32.71
50	0.005643	91,801	27.92	0.003227	95,354	31.8
51	0.006079	91,283	27.07	0.003476	95,046	30.9
52	0.006538	90,728	26.24	0.003763	94,716	30.01
53	0.007018	90,135	25.4	0.004091	94,360	29.12
54	0.007535	89,502	24.58	0.004465	93,974	28.24
55	0.008106	88,828	23.76	0.004884	93,554	27.36
56	0.008755	88,108	22.95	0.005349	93,097	26.5
57	0.0095	87,336	22.15	0.005861	92,599	25.64
58	0.010356	86,507	21.36	0.006423	92,056	24.78
59	0.01132	85,611	20.58	0.00704	91,465	23.94
60	0.012405	84,642	19.81	0.007732	90,821	23.11
61	0.013589	83,592	19.05	0.008497	90,119	22.28
62	0.01484	82,456	18.31	0.009318	89,353	21.47
63	0.016149	81,232	17.57	0.010192	88,521	20.67
64	0.017547	79,920	16.85	0.011138	87,618	19.88
65	0.019102	78,518	16.15	0.012199	86,642	19.09

[a] Probability of dying within one year.
[b] Number of survivors out of 100,000 born alive.

Taking the figures for a thirty-year-old, the numbers translate to 1.39 deaths per thousand men and 0.63 deaths per thousand women. Wow, that's pretty darned low. Moving to fifty-year-olds, we find 5.64 men per thousand and 3.23 women per thousand.

That's the great thing about actuarial tables. They seem to always predict a nice future, although when you look back at them, they actually predict that the odds are that you shouldn't be here any more. For even greater accuracy, you should consider a person's racial and ethnic background. Whites, blacks, Asians, Hispanics—each has a different expectation of longevity, but the specifics are beyond the scope of this book.

The odds are that you'd like the insurance company to be right; that is, you'll live out the term for which you bought the insurance. But you'd like to leave something to help out your family should you and the insurance company be wrong. Based on actuarial tables, insurance companies develop tables of **premiums** and **payoffs.** The following simple example of a death rate table will help to illustrate how this all works.

Death Rates for White Americans

Age	Death/1000	
	Men	**Women**
30–34	1.4	0.65
35–39	2.1	1.2
40–44	3.23	1.8
45–49	5.4	2.9
50–54	8.8	4.3
55–59	14.1	6.5

Say you are a thirty-nine-year-old male. According to the table, the likelihood of your not surviving the next year is 2.1/1000. That means that for every $2.10 you pay for insurance, the perfect payoff would be $1000, assuming that the insuror is a very efficient not-for-profit company, but term insurance isn't normally sold that way. To figure out the payoff for an investment of $100, we divide $1000 by 2.1 and then multiply the quotient by $100:

$$\text{Payoff} = \$100 \times (1000 \div 2.1) = \$47,619$$

That means that for about $210 for the year's premium, that thirty-nine-year-old male should be able to buy $100,000 of term insurance. Why does $210 look so familiar? Well, if you look back at the above table, it's 100 times the death rate. Pretty convenient, eh? What do you suppose a thirty-nine-year-old woman would pay for $100,000 insurance. If you said $120, you got it. And a fifty-five-year-old man would pay $1410 for the same coverage, versus only $650 for a fifty-five-year-old woman.

It's practically a magic formula, although in real life it wouldn't work. That's because no insurance company is that efficient, and most insurance companies are out to make a profit. Nevertheless, the model is a pretty good representation of how rates skyrocket with increasing age. I'm sure you've seen ads for insurance for seniors up to age eighty-five at "very affordable rates." How can the companies afford to do that? There are two parts to the answer to that question. First off, the payouts are in tens of thousands, not hundreds of thousands. Second, there usually is a three year wait (read that three years worth of paid premiums) before the insured's death will result in a payout. That's why it can be affordable.

Pop Quiz

Use the "Death Rates for White Americans" table to answer the following questions.

1. What should a thirty-year-old white woman expect a year's premium to be for a $200,000 term policy?

2. What should a forty-seven-year-old white male expect to pay a year for a $50,000 term policy?

3. How much term insurance should a forty-one-year-old white woman expect to buy for a monthly premium of $45?

Answers

1. $130

2. $270

3. $300,000; (remember, the yearly premium is 12×45, or $540).

Nonterm Life

Insurance salespersons rarely push term insurance (the cheapest) to young people. Rather, they try to sell **whole life** or **universal life** policies. These are combination products with some kind of savings plan attached. The selling point is that as your insurance policy ages, it accumulates monetary value. That's because of the attached savings program. You would actually accumulate more money by purchasing term life insurance and putting your savings into a money market account or mutual funds. Either of those stands to pay you a better rate of return than an insurance policy, but neither of those pays the insurance salesperson as much as a combination policy, so guess which one she wants you to buy.

Since nonterm policies contain different savings plans, there is no way for me to tell you how to calculate premium from death-rate or actuarial tables, but you can bet your boots that actuarial tables are still at the root of the calculations. To that, the insurer adds the amount you pay into savings, the company's expenses (including its advertising costs), the salesperson's profit, the stockholders' profit, and *presto change-o*, out comes a premium amount.

It has been pointed out that life expectancy these days is considerably longer than it was in the days of our great-grandparents. In fact, it has been projected that a person born in 2006 might have a life expectancy twenty years greater than someone born in 2001, just due to the medical advances made in the intervening five years. In fact, by the year 2050, a retirement age of eighty-five might be considered the norm. Ninety-one percent of all U.S. males make it past their fiftieth birthday, 85% make it to sixty, and 78% make it to sixty-five. It's just a matter of time, with continued medical advances, for most of our children to be making it to one hundred and beyond. The question they'll have to deal with then is when medicine has made it possible to live into their second hundred years, will quality of life make living that long worthwhile?

Other Risks

Term life insurance is a great statistical model, since the actuarial tables specify the risk that's involved and the amount of the final payoff is written into the terms of the contract. But there are other types of insurance that are not so easy to evaluate. That includes homeowner insurance, flood insurance, earthquake insurance, and automobile insurance. In each of those cases there is a maximum provision set, but the actual amount of payoff is quite likely to vary with the circumstances. In every case, however, never forget that it pays to shop around. A wise consumer is an informed one.

Automobile Insurance

There are many different forms of automobile insurance, as any car owner knows. There is liability insurance, which everyone should have for his or her own protection, and which is required by law in many states. Then there are collision, comprehensive, and medical (which is sometimes included with liability).

Unlike term life insurance which always ends with the expiration of the policy or the policy holder, automobile insurance rates need to account for the possibility of fraud. Anybody that you as a driver hit is liable to sue you (hence your insurance company) for an indeterminate amount. That amount may be for injuries dreamed up by an attorney, or for collision damage dreamed up by a creative body shop with or without the person's collusion (collusion damage! I like that!).

There are two rules of thumb to follow when purchasing automobile insurance. First, know that no insurance company nor court will grant you more in damages than the Kelly Blue Book's valuation of your car. That means that if your car has $3000 worth of damage and a Blue Book value of $1200, you are not going to receive more than $1200. For all practical purposes, what this means is that as your car ages, your need to carry collision insurance diminishes. Why pay an $800 per year premium for a car that's only worth $900? Do yourself a favor and put the money into the bank (or other investment). You'll soon have saved enough for a replacement car.

The second rule is to keep your deductible as high as you can afford. Very few insurers offer collision or comprehensive insurance with a zero deductible, but if they do, you can bet that it's at least $100 more expensive than it would be with a $100 deductible. Better yet is a $500 deductible. Of course, should you sustain damage in an accident, you'll be out of pocket $500. Still, think of the hundreds you'll save in premiums over the course of the years you're accident free. Again, bank that difference. I keep using the word "bank," because it's a part of the language, but, as we'll discuss in Chapter 13, it's not necessarily the best place to invest money.

Pop Quiz

1. Why are you better off buying term life insurance than whole life?

2. Why would the insurance salesperson try to convince you to buy a whole life policy?

3. What are the two rules to follow to save money and get the best value from your automobile insurance?

Answers

1. Whole life charges higher premiums because it combines a low interest paying savings account with the insurance policy. If you buy term insurance and bank the difference in premiums you'd accumulate more money quicker.

2. (S)he stands to make more money from the sales commission.

3. Take the highest deductible that you can afford to pay out of your own pocket, and as your car ages, drop collision insurance altogether.

13

Simple Investing

As mentioned earlier in this book, banks are a common and oft used place to put money, so why don't banks put their money into other banks? Well, to put it plainly and simply, there are better ways to make money. That's right, make money. Most people make their money by working, but some make it by investing. Investing is a way of having your money work for you to make more money. While you probably work somewhere between thirty-five and fifty hours a week, money works twenty-four hours a day, seven days a week, and never rests. Money works in two ways. The first is by earning interest. You should already know about interest. If by some chance you don't, go check out Chapter 6.

Bank interest rates vary, and have been as low as 0.25% and as high as 9% in my memory. Whatever the bank is paying at any given time, you can be sure that it's the lowest rate you can get anywhere, with the possible exception of an insurance policy. There's a reason for this. Money that you put into the bank is the safest it can be anywhere—including under your mattress. Funds in a savings bank or in a savings account in a commercial bank are insured by the Federal Deposit Insurance Corporation—a branch of the U.S. Treasury. Funds in a Savings and Loan bank (or S & L) are insured by another federal government agency, the Federal Savings and Loan Insurance Corporation. In either case, your savings are protected up to a maximum of $100,000. What if you have more than $100,000? Put the excess into a different bank.

The most important thing to remember about investments is that the return is directly proportional to the risk. That means the bigger the risk your money faces (of being lost) the greater the return. "What kind of a deal is that?" you may well ask. Well, since money in a savings account is at zero risk, the bank is paying you just enough interest to lure you into not keeping your money under the mattress. But they're not doing it for your benefit. Oh no, they're not altruistic! They want your dollars so that they can invest in something that will make them more money, such as real estate. The dollars you put into the savings account are used by the bank to give mortgages. The only benefits of keeping your savings in a bank are that they are safe, and **liquid**— meaning you can get the money out at a moment's notice, should you need it for some emergency.

Fixed Income Securities

A **fixed income security** is a type of investment in which you invest a certain amount of money and receive fixed payments at predetermined periods of time, usually monthly, quarterly, semiannually, or annually. Unlike a variable income security, such as a savings account, where rates change based on some condition such as short-term interest rates, the payments you'll receive on a fixed income security are known in advance. You don't have access to your principal, however, until the security reaches maturity (its full term). At that time your principal is returned to you. One form of a fixed rate security is a government bond, where an investment of $1000 at 5% might get you a payment of $50 per year, until the date of maturity. At that time, your $1000 would be returned. Because the return on your investment is guaranteed, the interest rate is usually not very high.

Certificates of Deposit

A **certificate of deposit (CD)** is a variant on the fixed income security. It is a note available in fixed amounts for fixed periods of time. The rate is usually two to four points higher than the savings account interest rate available. They come in amounts from $500 on up, in increments of $500 at some levels and $1000 at others. Generally, the shorter the term of the CD, the lower its interest rate. Now if you're saying to yourself "I must be giving up something," you're right. You are not giving up safety, however. The money invested in a CD is just as safe as any in an insured savings account. You are, however, giving up the liquidity of that money. If you put your dollars into a six-month CD, and have an emergency need for cash in three months, you'd best have that cash elsewhere. You cannot cash in the CD until its full term has expired. Also, you do not receive any payments until the CD reaches maturity. Then you get back your investment and all accrued interest.

CDs come in six-month, twelve-month, two-year, and five-year terms. When you buy a CD you are not only locking up your money for a fixed period of time. You are also betting that interest rates are not going to rise during that time and become higher than the rate you locked in when you bought your CD.

Yields of a CD purchased for $1000

Rate	6 Months	1 Year	APY
2.0%	$1010.10	$1020.20	2.020%
2.5%	$1012.58	$1025.31	2.531%
3.0%	$1015.11	$1030.45	3.045%
3.5%	$1017.65	$1035.62	3.562%
4.0%	$1020.20	$1040.51	4.081%
4.5%	$1022.75	$1046.02	4.602%
5.0%	$1025.31	$1051.27	5.127%
5.5%	$1027.88	$1056.54	5.654%
6.0%	$1030.45	$1061.83	6.183%
6.5%	$1033.03	$1067.15	6.715%
7.0%	$1035.62	$1072.50	7.250%
7.5%	$1038.21	$1077.88	7.788%
8.0%	$1040.81	$1083.28	8.328%
8.5%	$1043.41	$1088.71	8.871%
9.0%	$1046.02	$1094.16	9.416%
9.5%	$1048.64	$1099.65	9.965%
10.0%	$1051.26	$1105.16	10.516%

The figures in the table above were computed using a CD rate calculator available on line at http://www.bankrate.com/brm/calc/cdc/CertDeposit.asp

APY stands for **Annual Percentage Yield.** The Rate is the rate of interest, which is compounded daily. To find what the yield would be on a different multiple of $1000, just multiply the amount after six months or the amount after one year by that multiple. In other words, to find the yield of $5000 invested at 6% for one year, run your finger down the left column to 6% and then across to the one-year column, where you'll see that $1061.83 × 5 = $5309.15.

Stocks

A **stock** is a type of security that indicates ownership in a company, and represents a claim on part of the corporation's assets and income. Remember, at the start of this chapter I said that interest was one of two ways for money to earn money. This is the second one. Depending on the performance of a company's stock, the shareholder may receive periodic **dividends** (a sharing of earnings with the stockholders). The stock may also **appreciate** in value (go up). Such an appreciation is known as a capital gain, and is taxed at a different rate (usually lower) than regular income by the federal government.

There are two main "flavors" of stock, common and preferred. **Common stock** usually entitles the owner to vote at semiannual or annual shareholders' meetings, and to receive dividends. **Preferred stock** owners generally do not have the right to vote at shareholder meetings, but they receive preferential treatment in their claims on company assets and earnings. That is, they receive dividends before the common shareholders do. In the event of a company's going bankrupt, and being liquidated (having its assets sold off), the preferred stockholder has priority in being paid.

Stocks are also known as shares, or equity. If nothing else, the last paragraph should have alerted you to the danger of stocks. They can go up, but they can also go down. Money can be made either way. People who count on the stock market's always going up are known as **bulls.** Those who make money when it goes down are called **bears.** A bear makes money by selling stock (s)he doesn't own, betting that by the end of the day, when (s)he has to produce the stock certificate, (s)he can buy it for less money than it was sold for. This is known as **selling short.** Stocks are sold by a **stockbroker.** Every developed country in the world has one or more **stock markets,** with the largest being the New York Stock Exchange (NYSE). Stocks are traded (bought and sold) every weekday except certain holidays during prescribed hours. In the case of the NYSE, those hours are 9 A.M. to 4 P.M.

For the long term, there is no better place to invest money than in the stock market. Be aware, though, that stock prices can fluctuate widely. Stocks are not for the faint of heart, but it is true that over the long run, as the U.S. economy continues to expand, stocks continue to rise. Even those who lost their shirts in the infamous stock market crash of 1929 would have made back everything they lost and then some by 1937. As my parents used to say, "The stock market has room for bulls and room for bears, but there's no room for pigs."

Bonds

A **bond** differs from a stock in that it is a debt investment with which the investor (possibly you) loans money to a corporation or to the government for a specified period of time and at a specified interest rate. The entity doing the borrowing issues a certificate or bond to the investor stating the **coupon rate** (jargon for interest rate) that will be paid, and the maturity date (when the borrowed money is to be returned). The main types of bonds are the municipal bond, the Treasury bond, the Treasury note, the Treasury bill, the corporate bond, and the zero-coupon bond. The rate of return on a bond varies indirectly with its security. That is, the higher the interest rate, the lower the security of the investment. For that reason alone, corporate bonds tend to offer a higher interest rate than government bonds. It is essential that before you invest in a bond, or any other investment where your principal is at risk, you research the history of the company. Checking a bond's rating should help you to determine the likelihood of the issuer's **defaulting** (not paying you back).

Mutual Funds

According to the SEC (Securities and Exchange Commission), a **mutual fund** is "a company that brings together money from many different people and invests it in stocks, bonds, and other assets." In other words, a mutual fund is like a basket that contains an assortment of assets, including stocks and bonds. When you purchase shares of a mutual fund, you're buying a piece of the basket, but you do not actually own any of the assets that the mutual fund owns. While you may not own any of the assets, those assets are important to you, since the value of the fund depends on the value of the assets it holds. As those assets increase or decrease in value, so does the value of the shares of the fund.

Mutual funds offer two benefits over buying either stocks or bonds. The first of those is diversification. Rather than buying many shares of one company's stocks or bonds, you purchase a piece of many companies' assets. In fact, you are buying a piece of every asset in the fund. The second advantage is professional management. That means that someone who spends much more time than you could in analyzing financial markets decides where to invest your money.

There are two categories of mutual funds, open-end and closed-end. **Open-end funds** have no limit on their number of shares. If you wish to buy some shares in the fund, the fund creates new shares and sells them to you. **Closed-end funds** have a fixed number of shares, and in order to purchase a piece of the fund you must wait until existing shares become available. There are many more funds of the open-end category.

Mutual funds are classified in many ways by different institutions, but the industry standard when it comes to classification is Morningstar. Morningstar has two classification systems, the Morningstar Style Box™ and the Morningstar Categories. The box is shown in the figure that follows.

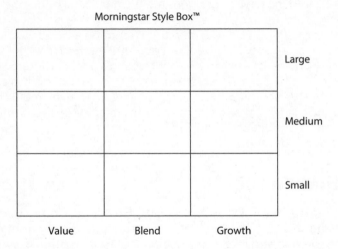

Morningstar Style Box™

Large

Medium

Small

Value   Blend   Growth

Equity stocks are categorized by the style and size of the equities they hold. Morningstar classifies fixed income mutual funds by duration and quality of the fixed income assets they hold. They also classify funds into forty-eight predetermined categories based on the assets they hold. These include Diversified Emerging Markets, Diversified Foreign, General Short Term, Latin America, Specialty Technology, and forty-three more. Check out their website for a complete list at www.morningstar.com.

Lipper (a division of Reuters) and Morningstar are global leaders in supplying information about and analysis of mutual funds. You need look no further than your local newspaper or some other source of quotes to see how a certain mutual fund is performing. By consulting either the Lipper Ratings or the Morningstar Ratings, you can see how one fund's performance compares to that of another. The SEC requires mutual funds to furnish histories so the prospective buyer can see how well a fund has performed over the course of the current year, one full year, three, five, and ten years, and since its inception. Besides the historical performance, both Lipper and Morningstar rate comparative performance of funds. Lipper rates funds using a numeric score with "1" being the highest and "5" being the lowest. Morningstar, not unreasonably, uses a star system with five stars being the highest and one star being the lowest.

Every mutual fund has expenses you should be aware of before investing in one, since some funds' expenses can have a dramatic effect on your returns. The three expenses you should be aware of are loads, redemption fees, and operating expenses. **Loads** are fees charged either when you buy (front-end load) or sell (back-end load) a mutual fund. They are usually used to pay a commission to the broker or agent from whom you buy the fund. Front-end loads are limited to 8.5% by law, but most fund companies charge less than that. Some companies sell no-load funds (so called because you pay no load), and some claim there is no reason to buy any other kind.

Redemption fees are used to discourage rapid turnover on funds. They are charges imposed if you sell a fund before a certain amount of time elapses. **Operating expenses** are a normal part of doing business, and include fees paid to the fund's manager for time and expertise, the cost of advertising the fund, and distribution expenses.

Although every mutual fund has expenses, some funds' expenses are very high, while others' are very low. A mutual fund company outlines everything you could possibly want to know about it in its **prospectus**—a booklet that identifies everything from the fund's objectives and past performance to the fees associated with it and background of the fund's manager. If there's anything you might want to know about the fund that you can't find in the prospectus, it provides you with the contact information you can use to get your questions answered. Unfortunately, many investors never read their funds' prospectuses. Don't be one of them.

There are many ways to buy mutual funds. You can invest through a company-sponsored 401(k) plan, through a personal IRA, or through a standard brokerage account, but you should be aware that mutual funds don't trade like stocks. You can only buy or sell them after the end of the stock market's hours. That's because mutual funds are traded based on their **NAV (net asset value).** In order to figure that out, the company looks at the values of all the assets in the basket, determines their total value, and divides that number by the total number of shares. Since this is a pretty complicated procedure, the company doesn't want to do it more than once a day.

To compensate for the inflexibility of only trading once a day, mutual funds permit investors to buy fractional shares. Suppose you had $200 to invest, and shares were selling for $52.50. If you could only buy whole shares, you'd only be able to buy three shares, and $42.50 would have to sit around until you had another $10 to invest. With fractional shares for sale, though, you can spend the entire $200 and buy 3.81 shares.

The Money Market

The money market is a part of the fixed income market, which we usually think of as synonymous with bonds, even though a bond is just one type of fixed income security. The main difference between the bond market and the money market is that the latter specializes in very short-term debt—the kind that matures in less than one year. Because of their short maturities, money market investments are also known as cash investments. They are in essence IOUs issued by large corporations, governments, and financial institutions. They are very liquid, and very safe, which is why they pay significantly lower rates than most other securities.

There is no central exchange or trading floor for money market securities, and most are in very high denominations, which limit access by the individual investor. The money market is a dealer market, where deals are transacted over the phone or electronically. Firms buy and sell securities in their own accounts at their own risk. That means there is no broker or agent to receive a commission.

The easiest way for individuals to gain access to the money market is with money market mutual funds or through a money market bank account. These pool the assets of thousands of investors in order to purchase money market certificates for them. Some money market instruments, such as Treasury bills, can be purchased directly.

Pop Quiz

1. What are the two main differences between a fixed income security and a stock?
2. What is the primary advantage of investing in mutual funds?
3. What are the two major global sources of information about and analysis of mutual funds?
4. What is the difference between stocks and bonds?

Answers

1. A fixed income security guarantees a return and therefore tends to pay a low rate of return. A stock is riskier and therefore tends to pay a higher rate of return if it pays any at all.

2. Mutual funds spread the risk around by holding multiple investment instruments at one time. While some of those may not do well, others are likely to do very well, and so the investor's money is safer than when invested in one stock only.

3. Lipper and Morningstar.

4. Stocks give the investor equity—a share in the worth of the company. Bonds are loans to the company and carry no ownership.

14

Distance Problems

Yes, this chapter is about distance problems, and its cornerstones are the following equations:

$$\text{rate} \times \text{time} = \text{distance}$$

and

$$\text{miles} \div \text{gallons} = \text{miles per gallon}$$

But let's start out with a more useful tool; one that will help you to solve a lot of relevant problems in real life. It's the "A" word, which we've made a point of avoiding until now. That's right, I mean algebra, but not the hard stuff, so don't panic. Unless you're an engineer you're never going to need quadratic equations, and if you are an engineer, you're probably reading this book for its entertainment value. Algebra is a tool for solving problems, and at its introductory level, it's very straightforward.

Algebra contains two types of numbers: the ones you already know, 1, 2, 3, . . . , called **constants**, and ones that are represented by letters, called **variables**. A variable behaves just like a constant, and it is worth the same thing in any given equation, but it can represent different numbers in different equations. Consider the equation $x = 3$. It means exactly what it says, namely a certain quantity, to which the name x has been assigned, is worth 3. But, some other time, we might decide to assign a different value to x, say $x = 7$. Well, no problem there, as long as you're consistent.

Now, there are four operations with algebra that serve us well in everyday problem solving. Those are addition, subtraction, multiplication, and division. Sound familiar? I certainly hope so. When constants and variables are combined by addition or subtraction, they look like this:

Addition: $x + 3$

Subtraction: $y - 4$

Multiplication and division look like this:

Multiplication: $2x$ or $2 \cdot x$ or $(2)(x)$ or $2(x)$ or $(2)x$

That's no sign, multiplication dot, or some form of use of parentheses.

Division: $\frac{y}{3}$ or $\frac{7}{z}$

Read that the first number over the second number, i.e., y over 3 or 7 over z.

The multiplication $2x$ gives rise to an interesting question. If that means two times the variable x, how would one write one times the variable x? In case you haven't guessed, x means one times the variable x. Think about it. If the **coefficient** (the number written immediately next to, hence the multiplier) of x were 0, there'd be nothing there, since 0 times anything is 0.

Think of a variable as an entity, such as a piece of fruit. With that in mind, what do you suppose is the sum of $x + x$? To answer that, think of the fruit analogy. What's 1 orange + 1 orange? Of course, it's 2 oranges. Therefore, $x + x = 2x$. For the same reason, $2x + x = 3x$, and $3z + z = 4z$. The rule we can infer from these examples is that when two numbers containing the same variable are added, the variable remains the same as would the piece of fruit. Only the numeric coefficients (the numbers next to the variables) are added. A similar rule applies to subtraction, with one exception:

$$4a - a = 3a, 5b - 2b = 3b, \text{ but } 4c - 4c = 0.$$

Solving Linear Equations

A **linear equation** is an equation that graphs as a straight line. An equation is any set of arithmetic or algebraic terms that are separated by an equals sign (=). Bear in mind that that sign means what it says.

$$x + 7 = 16$$

An equation may be thought of as a mathematical sentence. It expresses a complete thought. What is on one side of the "=" is exactly the same as what's on the other. It's just written differently. To solve an equation, we want the variable alone on one side of the "=" and a constant on the other. The way to achieve that is to undo whatever non-variable(s) on one side is (are) combined with that variable. In this case, there's a 7 added to the variable. We undo addition by subtraction, and **an equation's meaning is unchanged if the same procedure is applied to both sides of it**. To undo the addition of 7, we subtract 7 *from both sides.*

$$x + 7 = 16$$
$$x + 7 - 7 = 16 - 7$$
$$x = 9$$

Solve a subtraction the same way, by undoing it. To undo a subtraction, add.

$$z - 7 = 16$$
$$z - 7 + 7 = 16 + 7$$
$$z = 23$$

The way to undo a multiplication, I'm sure you'll recall, is by division, so

$$7x = 63$$
$$\frac{7x}{7} = \frac{63}{7}$$
$$x = 9$$

Finally, the way to undo a division is by multiplication, therefore

$$\frac{p}{5} = 21$$
$$5\left(\frac{p}{5}\right) = 5(21)$$
$$p = 105$$

One note on a situation you're not going to encounter in this book, but you might in real life. Suppose you have an equation that combines addition or subtraction with multiplication or division, such as

$$3x - 7 = 14$$

In such a case, do all the adding or subtracting first, so that all variables are on one side of the equation, and all constants on the other. Then multiply or divide.

$$3x - 7 = 14$$
$$3x - 7 + 7 = 14 + 7$$
$$3x = 2$$
$$\frac{3x}{3} = \frac{21}{3}$$
$$x = 7$$

Now try a few.

Pop Quiz
Solve each equation for the variable.

1. $5a = 45$
2. $b + 9 = 26$
3. $s - 8 = 21$
4. $\frac{t}{7} = 12$

Answers

1. $a = 9$
2. $b = 17$
3. $s = 29$
4. $r = 84$

Distance Problems for Real

You may be muttering under your breath at the way I began this chapter, but you're going to thank me shortly. As noted at the beginning of this chapter, distance = rate × time. Algebraically, we represent that using the first letters of each of the three items, or

$$d = r \cdot t$$

This formula is not restricted to just finding distance. If any two of the variables are known, we can solve to find the third. Check this out:

---

1. Reese takes an airplane trip that lasts six hours. The plane flew at an average speed of 450 miles per hour (hereafter mph). How far did Reese travel?

This problem is very straightforward, since it tells us the time and the rate (speed). All we need to do is to plug them into the formula and solve for distance.

$$d = r \cdot t$$
$$d = (450)(6)$$
$$d = 2700 \text{ miles}$$

---

2. Alex drives 250 miles and averages 50 mph. How long does the trip take him?

Again, we'll plug in the values.

$$d = r \cdot t$$
$$250 = 50 \cdot t$$
$$\frac{250}{50} = 50 \cdot \frac{t}{50}$$
$$5 = t$$
$$t = 5 \text{ hours}$$

Notice that in this solution, the variable was on the right side of the equation for most of the time. There is no wrong side for the variable to be on. Both sides are equal. Now try this one.

---

3. It takes Kira 8 hours to travel a distance of 240 miles. What was her average speed for the trip?

Once more, we'll plug in the values.

$$d = r \cdot t$$
$$240 = 8 \cdot r$$
$$\frac{240}{8} = 8 \cdot \frac{r}{8}$$
$$30 = r$$
$$r = 30 \text{ mph}$$

Is it starting to make sense? Hopefully you can see the practical value of distance problems in planning a trip. We'll get into some more complex, if still somewhat practical, distance problems in a bit. First let's look at another topic that's on today's drivers' minds, and that's fuel economy.

Fuel Economy

You should know what it's costing you to run your car, and while the old saying goes, an automobile engine runs on oil, since in fact, all moving engine parts are coated with that lubricant, in order to get anywhere, you're going to need gasoline. The higher the price of gasoline gets, the more concerned the average driver is with fuel economy, which is why in 2006, when I am writing this, the gas-guzzling SUVs (sport utility vehicles) are not selling very well. In my memory, gas has been as low as $0.26 per gallon and as high as $3.46 in my neck of the woods, which is near the New Jersey/Pennsylvania border. Now you can't do anything about the cost of gasoline, other than looking for the filling station with the best price, but you can monitor your vehicle's fuel efficiency. That might point you toward buying a more efficient vehicle next time, or it just might help you to know when your vehicle's engine needs servicing. If a vehicle has been giving you 26 miles per gallon, and suddenly that drops to 18.3, that's a sure sign that something is wrong.

Calculating fuel economy is a simple matter of division. Miles per gallon means miles ÷ gallons. Make it a habit to zero your trip odometer each time you fill up your gas tank. Then, on the following fill up, divide the number of miles on the odometer by the number of gallons that go into your tank. Do both to the nearer tenth. So, if you've driven 324.5 miles, and your tank accepts 14.2 gallons, then 324.5 ÷ 14.2 = 22.85 mpg. Based on the vehicle you're driving, you'll have to decide whether that's good or bad.

More Distance Problems

Not all distance problems are concerned with finding the distance. Sometimes you'll know the distance and wish to find the rate, while at others you'll want to find the time. Examine these.

4. Rocio drives due west at 45 mph. Myles starts out at the same place and at the same time, but drives due east at 55 mph. a) In how long will they be 200 miles apart? b) How far will each have driven?

To solve this problem, we have to consider what the elements of the problem are. No matter how far each of them drives, they'll drive a total of 200 miles, and they both will drive for the same amount of time, so the solution to part a is to be found by the equation that adds the distances they both drive:

$$rt_{Rocio} + rt_{Myles} = 200$$
$$45t + 55t = 200$$
$$100t = 200$$
$$t = 2 \text{ hours}$$

That's the answer to Question 4a. To find the answers to Question 4b, multiply each rate by the time to get the distance each drove. That's 90 miles for Rocio and 110 miles for Myles. Cool? Here's another.

---

5. One airplane flies due north at 300 miles per hour. A second starts from the same point and flies due east at 400 miles per hour. What is the distance between the two planes after two hours have passed?

Are you confused? You should be able to solve this problem in your head. Are you ready to do it or should I tell you the answer? It's 1000 miles. Since one plane is flying north and the other is flying east, their paths form a right angle. The distance between the planes will always be the hypotenuse of a right triangle, as you can see in the following figure.

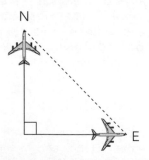

After two hours the north-heading plane will have flown 600 miles and the one heading east will have flown 800 miles (by use of $d = rt$). That's 200 times the legs of a 3-4-5 right triangle (if you need to, go back to Chapter 8 and check out "Pythagorean triples").

All right, I know that was a sneaky one, but I warned you much earlier that I would try to trick you from time to time. Here's a more straightforward one.

6. Hailee starts out from the bus depot at 5 P.M., heading south on Route 7 at 45 miles per hour. Frank starts out from the same place and going in the same direction at 6 P.M. driving at 50 mph. At what time will Frank catch up to Hailee?

For the solution, I'll arbitrarily let Frank's time = t, and Hailee's time be $t + 1$, since she will have driven one hour longer than he, having left an hour earlier. When Frank catches up, they will have both driven the same distance, so

$$rt = rt$$
$$45(t + 1) = 50t$$
$$45t + 45 = 50t$$
$$45t + 45 - 45t = 50t - 45t$$
$$45 = 5t$$
$$t = 9 \text{ hours}$$

If Frank's trip takes 9 hours, he'll catch up with her at 3 A.M. Now you try a few.

Pop Quiz
Solve each problem.

1. It takes Sebastian five hours to get to his destination traveling at 65 mph. How far is his destination?

2. Ali traveled 230 miles in four hours. What was her average speed?

3. Jake's car averaged 22 miles per gallon. After he had driven 350 miles he filled up his tank at $2.36 per gallon. What did it cost him?

4. Mariel flew 3000 miles from the east coast to the west coast in 7.5 hours. The trip back took her 6.75 hours. How much faster did her plane fly on the return trip?

Answers

1. 325 miles
2. 57.5 mph
3. $37.55
4. 44.44 mph

Measurement Tip Sheet

Measurement

International System of Units (SI)

SI Base Units		
Unit	*Symbol*	*Measured*
ampere	A	Electric current
kelvin*	K	temperature
gram	g	mass
liter	L	capacity
meter	m	length
seconds	s	time

0°K = -273°C, or "absolute" zero, the freezing point of nitrogen.

SI Prefixes

Prefixes are used before the base unit to indicate the multiplier, e.g. kilo before gram is abbreviated kg and means 1000 grams.

Prefixes exist beyond the range included here, but you are not likely to have a practical use for them.

Prefix	Symbol	Multiplier
mega-	M	1,000,000
kilo-	k	1,000
hecto-	h	100
deca-; deka	da or D	10
deci-	d	.1
centi-	c	.01
milli-	m	.001
micro-	μ	.0000001
nano-	n	.000000001

COMMON CONVERSIONS

1 kg	≈	2.2 lbs.
2.54 cm	≈	1 in.
1 m	≈	39.37 in.
1 L	≈	1.057 liquid quarts
1 km	≈	0.62 mi.

TRADITIONAL MEASURE

Length		
1 inch	1000 mils	2.54 cm
1 foot	12 inches	30.54 cm
1 yard	3 feet	0.914 m
1 fathom*	6 feet	1.83 m
1 rod	5.5 yards	5.03 m
1 furlong	40 rods	201.17 m
(statute) mile	8 furlongs; 5280 ft.; 1728 yards	1.61 km

* a nautical measure

Liquid Measure

1 minim	0.0038 in^3	0.0616 ml or mL
1 fluid dram	60 minims	3.697 ml or mL
1 fluid ounce	8 fluid drams	29.574 ml or mL
1 cup	8 fluid ounces	0.214 L
1 pint	2 cups	0.473 L
1 quart	2 pints	0.946 L
1 gallon	4 quarts	3.785 L

Avoirdupois Weight (avdp)

1 grain		64.799 mg
1 dram	27.344 grains	1.772 g
1 ounce	16 drams	28.35 g
1 pound	16 ounces	453.592 g
1 ton	2000 pounds	907.185 kg

Area

1 square inch		6.452 cm^2
1 square foot	144 in^2	929.03 cm^2
1 square yard	9 ft^2	0.836 m^2
1 square rod	30.25 yd^2	25.293 m^2
1 acre	160 square rods	765.11 m^2
1 square mile	640 acres	2.59 km^2

Note: All decimals have been rounded to the nearest thousandth.

B

Addition and Multiplication Flash Cards

These are the addition/multiplication flash cards I promised you in Chapter 3. You can use a copy machine to make copies of the cards, or you can cut them out and use them as they are (the answers are on the backs). Make sure you learn each fact backwards and forwards, since I've provided them only one way. That is, I've given you 4 (+/×) 8, but not 8 (+/×) 4. You need to recognize them both ways.

$1 + 2 =$	$1 + 3 =$	$1 + 4 =$
$1 + 5 =$	$1 + 6 =$	$1 + 7 =$
$1 + 8 =$	$1 + 9 =$	$2 + 2 =$

$1 + 4 = 5$	$1 + 3 = 4$	$1 + 2 = 3$
$1 + 7 = 8$	$1 + 6 = 7$	$1 + 5 = 6$
$2 + 2 = 4$	$1 + 9 = 10$	$1 + 8 = 9$

$2 + 3 =$	$2 + 4 =$	$2 + 5 =$
$2 + 6 =$	$2 + 7 =$	$2 + 8 =$
$2 + 9 =$	$3 + 3 =$	$3 + 4 =$

$2 + 5 = 7$

$2 + 4 = 6$

$2 + 3 = 5$

$2 + 8 = 10$

$2 + 7 = 9$

$2 + 6 = 8$

$3 + 4 = 7$

$3 + 3 = 6$

$2 + 9 = 11$

3 + 5 =	3 + 6 =	3 + 7 =
3 + 8 =	3 + 9 =	4 + 4 =
4 + 5 =	4 + 6 =	4 + 7 =

$3 + 7 = 10$	$3 + 6 = 9$	$3 + 5 = 8$
$4 + 4 = 8$	$3 + 9 = 12$	$3 + 8 = 11$
$4 + 7 = 11$	$4 + 6 = 10$	$4 + 5 = 9$

$4 + 8 =$	$4 + 9 =$	$5 + 5 =$
$5 + 6 =$	$5 + 7 =$	$5 + 8 =$
$5 + 9 =$	$6 + 6 =$	$6 + 7 =$

$5 + 5 = 10$	$4 + 9 = 13$	$4 + 8 = 12$
$5 + 8 = 13$	$5 + 7 = 12$	$5 + 6 = 11$
$6 + 7 = 13$	$6 + 6 = 12$	$5 + 9 = 14$

7 + 7 =	6 + 9 =	6 + 8 =
8 + 8 =	7 + 9 =	7 + 8 =
1 + 1 =	9 + 9 =	8 + 9 =

Addition Set 5: Back

$7 + 7 = 14$	$6 + 9 = 15$	$6 + 8 = 14$
$8 + 8 = 16$	$7 + 9 = 16$	$7 + 8 = 15$
$1 + 1 = 2$	$9 + 9 = 18$	$8 + 9 = 17$

$1 \times 4 =$	$1 \times 7 =$	$2 \times 2 =$
$1 \times 3 =$	$1 \times 6 =$	$1 \times 9 =$
$1 \times 2 =$	$1 \times 5 =$	$1 \times 8 =$

$1 \times 4 = 4$	$1 \times 3 = 3$	$1 \times 2 = 2$
$1 \times 7 = 7$	$1 \times 6 = 6$	$1 \times 5 = 5$
$2 \times 2 = 4$	$1 \times 9 = 9$	$1 \times 8 = 8$

$2 \times 3 =$	$2 \times 4 =$	$2 \times 5 =$
$2 \times 6 =$	$2 \times 7 =$	$2 \times 8 =$
$2 \times 9 =$	$3 \times 3 =$	$3 \times 4 =$

$2 \times 5 = 10$	$2 \times 4 = 8$	$2 \times 3 = 6$
$2 \times 8 = 16$	$2 \times 7 = 14$	$2 \times 6 = 12$
$3 \times 4 = 12$	$3 \times 3 = 9$	$2 \times 9 = 18$

$3 \times 5 =$	$3 \times 6 =$	$3 \times 7 =$
$3 \times 8 =$	$3 \times 9 =$	$4 \times 4 =$
$4 \times 5 =$	$4 \times 6 =$	$4 \times 7 =$

3 × 7 = 21	3 × 6 = 18	3 × 5 = 15
4 × 4 = 16	3 × 9 = 27	3 × 8 = 24
4 × 7 = 28	4 × 6 = 24	4 × 5 = 20

$4 \times 8 =$	$4 \times 9 =$	$5 \times 5 =$
$5 \times 6 =$	$5 \times 7 =$	$5 \times 8 =$
$5 \times 9 =$	$6 \times 6 =$	$6 \times 7 =$

$5 \times 5 = 25$	$4 \times 9 = 36$	$4 \times 8 = 32$
$5 \times 8 = 40$	$5 \times 7 = 35$	$5 \times 6 = 30$
$6 \times 7 = 42$	$6 \times 6 = 36$	$5 \times 9 = 45$

$6 \times 8 =$	$6 \times 9 =$	$7 \times 7 =$
$7 \times 8 =$	$7 \times 9 =$	$8 \times 8 =$
$8 \times 9 =$	$9 \times 9 =$	$1 \times 1 =$

$7 \times 7 = 49$

$6 \times 9 = 54$

$6 \times 8 = 48$

$8 \times 8 = 64$

$7 \times 9 = 63$

$7 \times 8 = 56$

$1 \times 1 = 1$

$9 \times 9 = 81$

$8 \times 9 = 72$

Calculating the Tip

Remember, to find 10% of any value, move the decimal point one place to the left. For instance: 10% of $43 is $4.30.

Taxi Cab

20% is the usual appropriate tip for a taxi cab driver.

It's easy to find 20% in your head. It's twice as much as 10%. Find 10% of the amount and double it, or use the table on the reverse side of this card.

Restaurant Advice

In **Europe** it is customary for the restaurant to include the gratuity in your bill, so don't tip on top of the built-in tip. In **U.S.** restaurants an average tip (for average service) should be about 15% of the check before tax. Since waitpersons make a substantial amount of their income from tips, not leaving one should only be done if that person's performance was awful. Never short the waitperson because you didn't care for the quality of the food. Let the restaurant manager or maitre d' know that.

For poor to very poor service, you might want to reduce the size of the tip to between 10 and 12% of the cost of the meal.

For good to very good service, you might tip between 18 and 20% of your total (before tax). Keep in mind that, as the person leaving the tip, **you decide** how much the service was worth.

If you're planning on returning to the restaurant, it's better to have a reputation for being an average to better-than-average tipper than a poor one. It's liable to get you better service next time.

Computing Tips

The table on the following page calculates tips on amounts in intervals of $10. If you'd like an exact percentage for an in-between amount, move the decimal point one place to the left. For example, to find 15% of $64:

1. Find 15% of $60 ($9.00).
2. Find 15% of $40 ($6.00).
3. Move the decimal point on the $6.00 left to get $.60.
4. 15% of $64 is $9.00 + $.60 = $9.60

All amounts in the table are in dollars and cents. For larger amounts, combine, For example:

To find 15% of $140 add (15% of $100) + (15% of $40).

	$10	$20	$30	$40	$50	$60	$70	$80	$90	$100
8%	.80	1.60	2.40	3.20	4	4.80	5.60	6.40	7.20	8
9%	.90	1.80	2.70	3.60	4.50	5.40	6.30	7.20	8.10	9
10%	1	2	3	4	5	6	7	8	9	10
11%	1.10	2.20	3.30	4.40	5.50	6.60	7.70	8.80	9.90	11
12%	1.20	2.40	3.60	4.80	6	7.20	8.40	9.60	10.80	12
13%	1.30	2.60	3.90	5.20	6.50	7.80	9.10	10.40	11.70	13
14%	1.40	2.80	4.20	5.60	7.00	8.40	9.80	11.20	12.60	14
15%	1.50	3	4.50	6	7.50	9	10.50	12	13.50	15
16%	1.60	3.20	4.80	6.40	8	9.60	11.20	12.80	14.40	16
17%	1.70	3.40	5.10	6.80	8.50	10.20	11.90	13.60	15.30	17
18%	1.80	3.60	5.40	7.20	9	10.80	12.60	14.40	16.20	18
19%	1.90	3.80	5.70	7.60	9.50	11.40	13.30	15.20	17.10	19
20%	2	4	6	8	10	12	14	16	18	20

Figuring Discounts and Sales Tax

(Discounted price) = (Original price) − [(Original price) × (Percent of discount)]

(Discounted price) = [100% − (Percent of discount)] × (Original price)

$$(\text{Original price}) = \frac{(\text{Discounted price})}{[100\% - (\text{Percent of discount})]}$$

$$(\text{Percent of Original price}) = \left[\frac{(\text{Sale price})}{(\text{Original price})}\right] \times 100$$

(Percent of discount) = 100% − (Percent of original price)

Any percentage based upon $100 is the same as the dollar amount.

To find 10% of a number move the decimal point one place left.

To find 1% of a number move the decimal point two places left.

5% of a number is 10% of that number ÷ 2.

20% of a number is 10% of that number × 2.

30% of a number is 10% of that number × 3, etc.

Computing Sales Tax

	$1	$2	$3	$4	$5	$6	$7	$8	$9	$10
1%	.01	.02	.03	.04	.05	.06	.07	.08	.09	.10
2%	.02	.04	.06	.08	.10	.12	.14	.16	.18	.20
3%	.03	.06	.08	.12	.15	.18	.21	.24	.27	.30
4%	.04	.08	.12	.16	.20	.24	.28	.32	.36	.40
5%	.05	.10	.15	.20	.25	.30	.35	.40	.45	.50
6%	.06	.12	.18	.24	.30	.36	.42	.48	.54	.60
7%	.07	.14	.21	.28	.35	.42	.49	.56	.63	.70
8%	.08	.16	.24	.32	.40	.48	.56	.64	.72	.80
9%	.09	.18	.27	.36	.45	.54	.63	.72	.81	.90

Index